Practical Builders' Estimating

FOURTH EDITION

W. Howard Wainwright and
A.A.B. Wood

Stanley Thornes (Publishers) Ltd

First published in 1967 by Hutchinson Education
Second edition 1970
Reprinted 1973, 1975
Third Edition 1977
Reprinted 1978
Fourth Edition 1981
Reprinted 1982, 1984, 1987, 1988

Reprinted 1991 by
Stanley Thornes (Publishers) Ltd
Old Station Drive
Leckhampton
CHELTENHAM GL53 0DN
England

British Library Cataloguing in Publication Data

Wainwright, Walter Howard
Practical builders' estimating—4th ed.
1. Building—Estimates
I. Title II. Wood, Adrian A B
629'.5 TH435

ISBN 0 7487 03993

Printed and bound in Great Britain by
Courier International Ltd., Tiptree, Essex

Practical Builders' Estimating

Contents

Preface

This book, although written primarily for building students, will also be of interest to the practising builders' estimators employed by the many thousands of small and medium-size contracting firms.

In the past, the training and education of builders' estimators has been very haphazard; hence the great diversity apparent in tenders. Also, many estimators are still very secretive about their methods.

The book contains notes and experience culled from many years of teaching and the running of successful practices as estimating consultants to a number of small and medium-size builders. The book attempts to show the student the methods of producing realistic unit rates with labour outputs expressed in tangible terms, which can be easily visualized. The system should enable the student gradually to develop a grasp of labour outputs which he will be able to apply to other items not featured in the book.

Many students have been dissatisfied with previous works which have tended to contain a great deal of factual data but very few examples of unit rate analysis. We have tried to give as many practical examples as possible, covering typical items met with in current construction practice, and at the same time to incorporate the relevant data within the analysis.

The book is aimed at students on courses validated by the Technicians Education Council, the Chartered Institute of Building, the Institute of Quantity Surveyors and the Royal Institution of Chartered Surveyors. The book is considered to be particularly appropriate for degree and diploma students in the polytechnics and universities.

This revised and up-dated edition incorporates modern practice and rates of labour and materials current at the beginning of 1981. In particular the text incorporates changes to conform to the sixth edition of *The Standard Method of Measurement of Building Works* and the *1980 JCT Standard Form of Building Contract.*

1 Estimating procedure

The building contract

Work connected with the construction of a new building, or with alterations and additions to an existing building, is invariably the subject of a contract agreement. The essence of such a contract is that the builder promises to erect the building as shown on the drawings and in accordance with the detailed specification in return for a certain sum of money, known as the *contract sum.*

A prospective building owner cannot give the builder instructions to erect the proposed building and hand him an open cheque to be made out when the building is completed. If this practice were to be adopted the result would be a cost plus account, with great advantage to the builder. Virtually all building contracts are obtained by competition between builders, and the subject of builders' estimating consists of the process involved in arriving at the contract sum. In recent years negotiated contracts have become more fashionable but the ability of both parties to be able to understand and appreciate the estimating process is the basis of such successful negotiation.

Contracts without quantities

Contracts of this nature are usually restricted to minor works, and the contract documents comprise the drawings, specification and form of agreement.

The drawings are the plans, elevations, sections, and large-scale details of the proposed building.

The specification is a document prepared by the architect to supplement the drawings. In the specification the mixes of concrete, types of bricks and quality of timber are described in detail, as well as the methods of work, such as how concrete is to be mixed and how brickwork is to be protected. All the essential information that will affect the price but cannot be written on the drawings will be mentioned in the specification.

The form of agreement is the legal document signed by both parties, which states that the builder contracts to erect the building in accordance with the drawings and specification and the client agrees for his part to pay the contract sum.

In the event of a dispute, the arbitrator must study the original drawings, specification and form of agreement, before he can arrive at any settlement. Hence the term *contract documents.*

Formerly, builders quoting for such contracts would carefully study the specification and the drawings and relying on past experience, would quote a lump-sum figure and be prepared to sign a contract with this figure as the contract. Then, as competition grew keener, builders could no longer rely on such crude methods of estimating, and found that they had to take off measurements, and prepare quantities of work involved; applying prices to these quantities, they arrived at a total estimate. The quantities they prepared were not only quantities of material, but specific elements of the building such as foundations, walls and roofs; they reduced these to units of square metres and applied their own unit prices, which included both material and labour costs.

With this type of contract, as the builder had to prepare his own quantities and prices, any errors of measurement that occurred were the builder's responsibility, since the detailed calculations did not form part of the contract.

Contracts with quantities

A contract which is based upon bills of quantities has the drawings, bill of quantities and form of agreement as the contract documents. The bill of quantities is usually prepared for the client by a professional quantity surveyor, whose fee is paid by the client. Several copies of the bill of quantities are prepared and sent out to the builders who are tendering for the work, to assist them with their estimating. The bill of quantities specifies the material and method of work to be used as well as the quantities of work to be done – in fact all the information that will affect the price. Any errors that may occur in the preparation of the bill of quantities will be rectified in the final account, so that the risk of errors by builders when preparing their quotations is considerably reduced.

When the contract is signed the bill of quantities and prices becomes part of the contract agreement and will be used in the preparation of the final account and in the settlement of any variations. The provision of bills of quantities leads to accurate tendering, as all the tenderers have identical documents from which to work and considerably reduces the costs of estimating for building work.

The bill of quantities

The bill of quantities is divided into sections, which are each based upon traditional trades. Each section is subdivided for easy reference. A typical bill of quantities would be divided up as follows:

1 Preliminaries
2 Excavation and earthwork
3 Concrete work
4 Brickwork and blockwork
5 Drainage
6 Roofing

7 Woodwork
8 Plasterwork and other finishes
9 Plumbing installations
10 Painting and decorating
11 Prime cost and provisional sums

As stated previously, the information supplied in a bill of quantities is guaranteed accurate to the builder; any errors that may occur in descriptions and/or quantities, or the omission of an item, will be treated as a variation and will be adjusted in the final account. The right under the contract to rectify errors does not, however, extend to the builder's errors in the computation of the contract sum. The quantity surveyor will check a builder's priced bill of quantities very carefully both arithmetically and technically before recommending the parties to sign a contract based upon the priced bills; the builder's attention will be drawn to any errors and he will be asked if he wishes either to amend the errors or to stand by the original offer. Once the contract is signed, the contract bills and all the rates, cash extensions and totals are considered accepted by both parties. The quantity surveyor will make every effort to check the builder's computations, but the builder would not succeed in any action for professional negligence against the surveyor in the event of errors of computation occurring in the contract bills. Unit rates and computations are the sole responsibility of the builder.

All bills of quantities prepared for contracts based upon the Standard Form of Building Contract are prepared in accordance with the rules of the Standard Method of Measurement of Building Work. This document, drawn up between the Royal Institution of Chartered Surveyors, the National Federation of Building Trades Employers, and others, provides a uniform basis for the measurement of building works and embodies the essentials of good practice. The Standard Method of Measurement has many advantages for the builder, who knows that each element of a building will be measured in a particular manner and that each element will be presented in the bill of quantities in a standard unit of measurement.

Methods of tendering

Open tendering

This method is often used to obtain tenders for building work. The prospective employer advertises in the national and technical press, giving brief details of the proposed works, and issues an open invitation to contractors to apply to him or his architect for the necessary documents. One of the conditions in the advertisement might be that the tenderer must pay a deposit which will be returned by the prospective employer on receipt of a *bona fide* tender. This provision is made to deter persons applying for documents out of mere curiosity.

It is usually stated in the advertisement and always in the tender documents that the employer does not bind himself to accept the lowest, or, indeed, any tender. The advertisement does not legally bind the employer in any way, but is merely an invitation to persons to make an offer. An offer must be unconditionally accepted before a contract can be made.

Indiscriminate requests for open tenders are generally unprofitable, since they often lead to building of poor quality, and the preparation of such tenders demands of the industry an inordinate expenditure of time, effort and money. Under this system, contractors are sometimes awarded work which they are ill-equipped to carry out, financially and practically. Although it may be pointed out that no special tender need be accepted, a committee spending public money is naturally tempted to accept the lowest offer.

Selective tendering

Under this method, competitive tenders are obtained by drawing up a short list of contractors and inviting them to submit quotations. This short list can be drawn up in two ways. Either the employer's professional advisers name suitable contractors, or an advertisement is put in the national and technical press, setting out brief details of the proposed scheme and requesting contractors who wish to be considered for inclusion in the short list to apply. The object of the selection is to make a list of firms, any one of which could be entrusted with the contract. The contractor chosen will simply be the one of these who submits the lowest tender.

The short list may be limited to the small number of firms who satisfy all the requirements. When this is not possible, guidance on the maximum number of tenderers is given in the Code of Procedure for Selective Tendering compiled by the National Joint Consultative Committee of Architects, Quantity Surveyors and Builders, and the Department of the Environment.

Negotiated contracts

This method of tendering is usually used for building work of a very difficult nature, where the magnitude of the contract may be unknown at first, or where early completion is most important, or where continuation or repetition of an existing contract is considered desirable. Under a contract of this nature there is usually not enough time to wait for drawings and bills of quantities to be prepared. There is always the danger that a price obtained in this way may be higher than one obtained by competition: however, intelligent use of a negotiated price may even result in a lower cost.

Jobs with difficult phasing programmes are the obvious cases where a negotiated price should be considered, but this method has other uses. By negotiating contracts it is possible to retain the services of firms which have been found satisfactory in the past.

Another method of negotiating contracts is the serial system, which can result in very keen prices. Here selected contractors are invited to tender for one building – a school, for example – on the understanding that the successful tenderer will be asked to build several other schools at the agreed rates. Naturally, this encourages keener prices than would the project for a single school, since the contractor will be able to organize his men and materials on the basis of an expected programme and carry out this work more efficiently. Furthermore, experience gained on the earlier jobs will be useful later on, as long as the contracts are basically similar. Although the serial type of negotiated contract can be based upon accurate bills of quantities, many negotiated contracts are arranged with one contractor on a prime cost basis. There are three types of prime cost contracts in current use: the *Cost plus, Cost plus fixed fee* and *Target cost with fluctuating fee.* The total cost of the work is recorded with these systems, and the contractor's fee is charged in addition to the cost in accordance with one of the three methods. Unfortunately rapid inflation in recent years has tended to curtail this type of contract.

The preparation of the tender

The tender documents usually consist of a letter of instructions to tenderers, the printed form of tender together with an endorsed envelope, two copies of the bills of quantities, and the necessary drawings and details showing the layout and elevations of the proposed building. The extent of the drawings will be governed by the rules imposed by SMM6, in particular, clauses A.2, A.5, B.3 and the references to drawn information required for each of the remaining sections in that document. The estimator studies the documents, lists all prime cost and provisional sums, and marks the quantities which the firm normally sub-lets. After carefully studying the drawings and the bills of quantities, an estimator of experience will have an approximate idea of the cost of the building. He then visits the site, accompanied by other members of the firm's management, where they discuss the project. They consider in particular possible difficulties of access to the site; the storage of materials (which is very important on confined city sites); adjoining premises; type of ground; necessity of temporary roads, hoardings, compounds and enclosures; the arrangement and siting of temporary offices and welfare buildings; the availability of water for construction; and the availability of labour in the area. The problem of security of the works against theft and vandalism must also be discussed as this is now an extremely expensive element.

At this stage a decision must be made whether to proceed with the tender any further; the decision will be made after the study of several other factors, such as the financing of the job, the proportion of the firm's own work to sub-let work, the time required to complete the contract, and known future commitments.

After the decision to proceed, the estimator extracts the sub-contractors' work from a copy of the bills of quantities and when these have been photocopied they are dispatched to the sub-contractors with a covering letter setting out all the relevant information about the contract. Pages from the preliminaries section of the bills of quantities and all preamble clauses pertaining to sections which affect the sub-contractors are copied and forwarded with the actual quantities, together with relevant drawings and details. Very often three or four prices are obtained for each section of the sub-let work. It is good practice to always keep one copy of the complete bill of quantities in the office, as this is extremely useful for reference when dealing with queries, and also helps to reduce the risk of errors.

The estimator then looks through the bills of quantities, making careful notes of all the materials for which he will require special quotations. Quotations are not normally required for every material; the prices for such commodities as bricks, cement and aggregates, which are used every day on most contracts, will be known. However, there are contracts where quotations are required for even these everyday materials; this would occur, for example, when preparing a tender for a building to be erected outside the normal area of operation of the company. Another case would be a tender for a building of such magnitude that generous trading discounts could be obtained because of the quantities of materials to be ordered.

The estimator then begins work on the analysis of the unit rates, for which no information is required and gradually, as the quotations for materials are returned, the unit rates are analysed so that, several days before the date for the submission of the tender, the estimator should have the rate analysis work completed. This work is usually carried out in pencil to allow easy alteration. When material quotations are received they should be carefully scrutinized, as very often delivery dates are more important than the cheapest price. When all the items are priced, the bills of quantities are extended and totalled by comptometers, sub-contractors' quotations which are likely to be accepted if the tender is successful are inserted and a total estimate for the contract is thus obtained.

When this stage of the work is completed the estimator will present his estimate to the directors of the company who will convert it into a tender. This should always be the final responsibility of the managing director. It is useful for the estimator to present his estimate in the form of a schedule, giving the managing director all the relevant information at a glance. A typical layout for such presentation is as follows:

Tender analysis

Tender for:	*Gross labour rates:*
Date due:	*Craftsman:*
Date of completion:	*Labourer:*

	£
Net value of firm's own work:	
Preliminaries:	
Labour:	
Materials:	
Plant:	
Add for overheads:	
Add for profit:	
Approved sub-contractors:	
Add for profit:	
Nominated sub-contractors:	
Nominated suppliers:	
Provisional sums, contingencies, etc.:	
Total of estimate £	

From this summary sheet the managing director will prepare the final tender figure for submission, which will be entered on the form of tender.

The directors will be influenced by their current work load as they make their final decision on the tender total. In 1973 it was common for most builders to be inundated with tender inquiries. Frequently a contractor considering his final tender would decide that as his order book was overflowing he would add an 'extra' profit margin to his estimates of, say 25 per cent, and still end up with the lowest tender. A different picture had emerged by 1977, with directors of many building firms forced into making unenviable decisions to cut their tenders by reducing net profit as they desperately tried to stay in business and keep their organization intact for better days. The workload in the building industry improved marginally in the last part of the decade but at the same time numerous firms, some well-known and established for many years, went into liquidation. At the beginning of 1981, it would seem that the prospects of higher levels of new construction work in the United Kingdom are bleak indeed. Generally, new buildings are no longer 'value for money' in the present economic climate.

The form of tender is signed by the director or the company secretary, placed in the endorsed envelope and dispatched either by post or by hand to the architect's office before the final date for the receipt of tenders. The priced bill of quantities should remain in the possession of the company and in the event of the tender being accepted, the unit rates, extensions and totals are copied on to two blank bills of quantities provided by the architect, one copy of which will become part of the contract documents.

When a tender is prepared for a contract without quantities, a very similar procedure is followed, but the surveyor or estimator of the company has to take off his own quantities of work before unit rates can be analysed and a tender figure obtained. Usually sub-contractors take off their own quantities from the drawings, but there are occasions when the surveyor or estimator takes off the quantities for the sub-contractors to price.

The unit rates

Every effort should be made to include in the unit rates all elements which are constant in a particular component when erected on any building site. Costs which are peculiar to one particular site should not be included in the unit rates; if these are eliminated and included elsewhere in the tender the estimator can then calculate basic unit rates for the elements which should vary only slightly from the usual ones. For example, if special travelling time and expenses can be eliminated from the unit rate for brickwork, the rate should be the same for two similar buildings, of which one is situated next to the firm's headquarters and the other ten kilometres away.

The reason for this approach to pricing is that the estimator quickly gets to know the basic unit rate for each element, which with experience, he can vary a little to suit the site conditions, without having to re-analyse from first principles every item on each tender. If, for example, the estimator is pricing brickwork in a tender, and is of the opinion that a bricklayer's output will fall from the fifty bricks an hour included in the basic unit rate to forty bricks an hour, then the cost of the fall in output can be calculated and added to the basic unit rate. If, however, the labour output and brick supply price change, then the estimator will analyse a new rate from basic principles.

A typical basic unit rate for a square metre of brickwork would include the following:

1 Cost of bricks
2 Waste on bricks
3 Mortar including mixer
4 Waste on mortar
5 Cost of bricklayer's time at the all-in hourly rate
6 Cost of attending labourer's time at the all-in hourly rate

General overheads and establishment charges

If building firms submitted tenders based upon unit rates calculated to cover net site costs they would lose financially; the hidden costs of managing the business would show as a loss in their accounts. These costs are known as general overheads or establishment charges and are extremely difficult to calculate accurately for each individual contract. The total sum of money expended in these overheads can be calculated from the previous financial year's accounts and expressed as a percentage of the turnover for that year. This percentage can be used to calculate the approximate overhead charges on estimates being prepared for future contracts. Below is a list of some of the items which should be included in the general overhead and establishment charges:

Directors' fees (managing directors)
Salaries (agents, surveyors, head office staff)
Wages (materials and wages clerks, and non-working foremen)
National Health Insurance (employer's contributions for non-productive staff only)
Private Superannuation Scheme contributions (employer's contributions for non-productive staff only)
Office rent, rates, insurances, etc.
Builder's yard rent, etc.
Electricity and heating costs for the office
Motor-car expenses (including depreciation, repairs, tax, insurance, fuel and oil on cars provided for head office staff)
Printing and stationery
Advertising
Postage and telephones
Furniture and fixtures (renewals and depreciation)
Bank charges, including interest on overdraft
Accountant's fees

Profit

All commercial undertakings are in business to make a profit. The small building firm with a sole proprietor expects to make a profit for the proprietor over and above the annual sum that the business pays him for his services. Private limited companies expect to make a profit for the directors over and above their directors' fees. Public *limited liability* companies expect to make a profit which will enable the firm to pay a dividend to the company's shareholders, in return for the use of their capital invested.

When estimates are prepared for the submission of a tender the estimated profit required must be included. Profit margins are usually expressed as a percentage of the money to be expected to finance a building contract. No real guide can be given for a profit percentage, as this figure varies considerably. Some of the larger national firms of building and civil engineering contractors work on a profit margin as low as 1½ to 2 per cent, but on a contract of £2 million in value this is a considerable sum of money. Many smaller businesses working on small works and difficult alteration schemes require a profit of 30 to 50 per cent to cover possible risks and to stay in business. A total figure of between 15 and 20 per cent in respect of overheads, establishment charges and profit would be considered as reasonable but this is dependent on how a particular contractor prices his work.

The profit to be included can be influenced by many factors, which are collectively referred to as the *market conditions.* Below is a short-list of the factors that must be considered before assessing the profit:

1 Quantity of work in hand, ranging from minimal to overloaded

2 Future commitments
3 Placing in previous tendering
4 Possibility of future work from the same source
5 Assessment of the degree of competition

Tenders are usually prepared on the basis of *net pricing* or *gross pricing*.

Net pricing

Under net pricing, the net site cost only is included when building up a basic unit rate for an item; the quantities are then extended at these rates and totalled, and a final net site cost is obtained for the complete building. The profit and general overheads elements are then calculated as lump sums which are based upon the total net site cost. The sum total of the net site cost and the general overheads and profit will give the tender figure.

Gross pricing

Under gross pricing, net site cost unit rates are calculated and a percentage is added to each unit rate to cover general overheads and profit. These gross rates are then inserted against the items in the bills of quantities, which are extended and totalled to give the tender figure.

Note: The pricing examples illustrated in this book are based on net pricing, thus avoiding the constant arithmetical repetition of adding a percentage to each rate to cover head office overheads, establishment charges and profit, etc. It is important, however, to include this factor in practice in a tender.

Advantages and disadvantages of net and gross pricing

One of the advantages of gross pricing is that when variations occur on contracts they are priced at the rates inserted in the bill of quantities, and therefore variations for additional work will automatically increase the amount included for general overheads and profit.

One disadvantage of gross pricing is that there is a greater margin of error when a percentage is added to each unit rate to cover profit and overheads, as the total of these two amounts is naturally 'rounded off'. If it were necessary to add one percentage for overheads and another for profit, the estimator's task would be intolerable when using the gross pricing method.

One of the advantages of the net pricing method is that the calculation of overheads and profit can be reduced to one or two operations, thus giving a more accurate tender, and if necessary separate sums can be calculated and included for overheads and profit. A disadvantage of net pricing is that automatic increases in the amount included for general overheads and profit do not take effect if variations or additions are made. The reason for this is that the preliminaries section is often chosen as the most suitable place for the sums included for profit and overheads. Clause 13.5 of the 1980 JCT Form of Contract now makes specific provision for such costs to be reimbursed to the builder whether the variation is of a minor or major nature. The same rates of payment would not apply if the variations were so extensive that the contract sum was considerably increased, or if the contract period had to be extended due to circumstances beyond the builder's control. In cases such as these, each individual item priced in the preliminaries section is considered in detail and is increased if appropriate.

When using the net pricing method this difficulty can be overcome as follows: when the lump sums to cover general overheads and profit have been calculated the whole of the bill of quantities can be re-priced with new unit rates, each of which carries a proportion of the general overhead and profit amounts. A certain amount of adjustment is entailed to ensure that the arithmetical calculations with the new rates correctly total the contract sum. The process of adjusting individual unit rates to fit the tender amount causes a great deal of clerical work, but such calculations need only be made in the event of the tender being accepted.

A builder will price his bills of quantities at net unit rates, add the calculated sums for general overheads and profit, and complete the tender figure on the form of tender. If the tender is not accepted, the bills are never seen and remain priced on the net basis. If it is successful, the adjusted unit rates can be made on a blank bill of quantities, thus leaving the original calculations untouched. When a builder knows that his tender is being considered, this extra work is worth while.

In the case of tenders prepared for contracts based upon drawings and specification only, re-calculation is unnecessary, as the tenderer's unit prices are never disclosed. However, in such contracts it is sometimes necessary for the builder to supply a schedule of unit rates of the major items involved, for use if subsequent variations have to be made. The work involved here is comparatively easy, as the unit rates have to be supplied for only a few items. Care should be taken to ensure that each unit rate carries a proportion of the general overheads and profit.

Separate pricing of labour and materials

When pricing bills of quantities, many estimators prefer to break down each unit rate into labour, materials, and plant, thus arriving at a sum for labour, materials and plant for each item in the bills of quantities. The sum total of each of these elements for the contract as a whole can be obtained by running four separate cash columns, the fourth column being reserved for the quantities extended at the full rate.

The format of the modern bill of quantities does not help the estimator who wishes to prepare his estimates in this manner. The most common method used to price a bill of quantities this way is to rule three additional columns through the printed descriptions, and to insert the rates in the spaces between them. This makes the bills of quantities very difficult to read, but as the information is only used within the firm, many estimators are content to use this method.

A far easier way to obtain the same result is to unclip the pages of the bills and paste each page on to a sheet of double foolscap paper which is divided into vertical columns. This can be done quite easily, as most bills of quantities are printed on one side of the paper and the columns on the opposite side show each element of the unit rate clearly. An example of this method is shown below. If the gross unit rate system of pricing is used, two further columns can be provided: one for general overheads and the other for profit.

One of the main reasons for the use of such a method of pricing is the fact that accurate estimates of labour, materials, and plant costs on a proposed scheme are extremely useful when calculating lump sums for insurances and other emoluments which are based upon the labour costs. Another advantage of this system is that the labour costs involved for each item can be seen at a glance, which is invaluable when drawing up bonus schemes for operatives.

Materials

The costs of materials vary considerably depending upon the firm of suppliers and the purchaser. Many other factors determine the cost of material; for example, the quantity to be purchased, the annual turnover of the supplier and the speed with which the purchaser settles his accounts. It follows that a builder who is a purchaser of large quantities of materials and a prompt payer of accounts will obtain more advantageous quotations than a builder of the opposite type.

As stated previously, some materials which are in constant use do not require special quotations, but if such materials are required in large quantities, or are to be delivered to a site outside the normal supply area, special quotations will be required. Estimators should always obtain quotations from suppliers for materials which they do not often use and should not rely on price lists for this information. A further reason for variations in the prices of materials is the differences in distance of haulage, carriage costs and the size of the loads required. For example, ready-mixed

	Concrete 1:3:6 mix 38 mm aggregate as described	
A	In wall foundations	20 m^3
B	In beds not exceeding 100 mm thick	12 m^3

concrete costs considerably more when delivered in 1-cubic-metre loads than in 3-cubic-metre loads.

Wastage of materials
Material waste must be reflected in the unit prices, since under the Standard Method of Measurement of Building Works all quantities must be given net, as they will appear in the completed building. Wastage allowances can only be assessed from past experience and observations of wastage should be constantly noted by the estimator. Wastage of materials varies from firm to firm and depends to a great extent on the skill of the buyer, and the efficiency of site foremen and the firm in general. For example, a building firm which allows concrete for a housing contract to be mixed in front of each pair of houses will have higher materials wastage from numerous aggregate piles than will the firm which mixes concrete in one central position on the site.

Materials wastage can be classified into six broad categories, as follows:

1 *Cutting waste* when sheet materials have to be cut for a specific component, for example, plywoods, blockboards, plasterboard, felt
2 *Application waste* occurs when most wet building materials such as plaster and other finishings are used; the term can be extended to wastage on many other materials such as bricks, tiles and timbers cut to length
3 *Stockpile waste* when most loose materials are dispersed on the site because of partial use, for example, aggregates, sand
4 *Residue waste* occurs with paints, glues and other materials which are normally delivered in containers, and are never completely used
5 *Transit waste* occurs with brittle materials which break in transit such as slates and tiles; in many cases such materials are dispatched on the understanding that breakages in transit are the purchaser's responsibility
6 *Theft and vandalism*

In addition to normal waste an allowance for laps should always be included in the unit rates for many building materials, but such allowances cannot be strictly described as waste, for example laps and passings in damp-proof courses, roof felts, lead flashings and tongue allowances in tongued and grooved flooring. The cost of unloading materials, placing in a storage position on site and further site transportation by hand or machine must also be carefully considered.

	Labour (£)		*Materials* (£)		*Plant* (£)	*Rate* (£)	*Total* (£)
£10.50	210.00	£17.00	340.00	£1.00	20.00	28.50	570.00
£13.00	156.00	£17.00	204.00	£1.00	12.00	31.00	372.00

The actual cost of labour

The actual cost of labour to the builder entails far more than the trade union minimum rate, or even the weekly gross wage shown on the wages sheet.

The additional costs over and above the basic trade union guaranteed minimum weekly earnings can be summarized under the following headings:

1 Guaranteed week and inclement weather (employer's payments for time in attendance without production)
2 Sick pay (supplemental to statutory entitlements)
3 Tool money
4 Training levy
5 National Insurance and state superannuation (employer's contribution)
6 Annual and public holidays with pay credits (employer's contribution)
7 Severance and redundancy payments

It is also convenient to make allowances in an 'all-in' hourly rate to cover the undermentioned factors:

8 Employer's liability and third party insurances
9 Trade supervision
10 Additional bonus payments
11 Travelling expenses
12 Non-productive overtime

Some estimators prefer to use the tràde union rates only in their analysis of the unit rate and then to calculate lump sums for each of the other emoluments when the bills are completely priced, inserting the lump sums in the preliminaries section of the bills of quantities.

This can be done quite simply if the bills of quantities have been priced with separate columns for labour, materials and plant as previously described. If this method is not used a very rough estimate can be obtained by dividing the total cost into 45 per cent labour and 55 per cent materials, but this is a very crude method of assessing these values, as the allocation percentages of labour and materials vary from trade to trade and from contract to contract. A far more satisfactory method is to convert these labour emoluments to an hourly amount which is added to the net trade union rate to give an all-in hourly rate.

The calculation of a typical 'all-in' rate expected at the beginning of 1981 is illustrated to indicate the method and principles involved but because of the constant changes in statutory legislation, Working Rule Agreements and wage rates, the figures become outdated very quickly. Several national journals and periodicals publish frequent up-to-date rates and it is advisable to consult these regularly.

In practice more exact 'all-in' rates should be built up based on an annual calculation in accordance with the principles as set out in the Code of Estimating Practice published by the Chartered Institute of Building.

		Craftsman (£)	**Labourer** (£)
1	Guaranteed minimum weekly earnings (plus 20p in London and Liverpool)	80.400	68.600
2	Inclement weather allowance – 2% –	1.608	1.372
3	Sick pay	0.400	0.350
4	Tool money (average)	0.500	—
5	Training levy	0.700	0.120
6	National Insurance and pension	12.500	10.700
7	Annual holiday credits and public holiday allowances	10.560	10.180
8	Death benefit scheme	0.100	0.100
9	Severance and redundancy payments, etc.	1.600	1.500
10	Trade supervision	6.800	5.800
11	Additional bonus payments	20.000	15.000
12	Travelling expenses	5.000	5.000
13	Overtime including non-productive element*	12.563	10.719
		152.731	129.441
14	Employer's liability and third party insurance add £2.00 per £100	3.055	2.589
	Weekly cost per operative	155.786	132.030
	Hourly cost based on a 45-hour week	£3.462	£2.934

Guaranteed time With effect from February 1970 the normal forty-hour week was fully guaranteed in accordance with the National Working Rule Agreement for the Building Industry.

The amount of lost time paid for by the employer can vary considerably from trade to trade and from year to year. The winter of 1962–3 provides an illustration of how much time can be lost; due to the severe weather, all work came to a halt for ten weeks. During this winter many firms 'laid off'

*An overtime allowance of one hour per day (excluding Saturday) has been included, but if this is not worked the totals would have to be divided by forty instead of forty-five hours. The hourly rates shown above should be reasonably realistic until new wage settlements are negotiated, probably in mid 1981. However, the factor which causes the biggest variation between different employers and different localities is in item 11 – additional bonus payments – and will be discussed later in this chapter.

Because of variable bonus payments and future likely wage settlements, the hourly rates used in this book have been rounded up to £5.00 for craftsmen and £4.00 for labourers.

thousands of operatives, but in many cases they had first paid a guaranteed weekly wage to every man for two or three weeks, in the hope that the weather would break. Such winters as this can throw calculations awry, but a fairly good average is to allow one hour, or 2 per cent per week per operative for time paid for and lost through bad weather. This should be included for all trades, irrespective of whether work is generally conducted indoors or outdoors. This unproductive time must be taken into account in the analysis of unit rates and one method of dealing with this problem is to divide the total all-in weekly cost of labour by thirty-nine hours rather than the guaranteed week of forty hours.

Sick pay The National Working Rule Agreement for the Building Industry lays down a set of rules whereby limited payments are made to an operative absent due to sickness or injury. Many private insurance schemes are in operation to cover this sick pay and the cost of the premium has been included in the 'all-in' rate.

Tool money The National Working Rule Agreement for the Building Industry provides for tool allowances to be paid to craftsmen and apprentices when they are in possession of a prescribed detailed list of tools. These allowances are all paid on a weekly basis, and the estimator should include this tool money allowance where applicable in the all-in hourly rate for every operative included in the analysis. As the allowances vary from trade to trade, an average amount should be taken.

Training levy The Construction Industry Training Board imposes a levy on most building firms based upon the number of workpeople in employment on a certain date. If firms take advantage of recognized training schemes they will receive grants as laid down by the Board. These grants should be credited to the training account in the general overheads. The amounts of these levies and grants vary from year to year.

National Insurance National Insurance is the system of collection of employer and employee contributions and, under the Social Security Act of 1975, supercedes the former stamp system and graduated pension contribution. An employer has to pay a weekly National Insurance contribution for each of his employees. The employee also pays weekly contributions and this is deducted by the employer from the employee's gross weekly wage. A record is kept of each deduction. When an employee changes his employment he takes a P.45 form, which shows the total National Insurance contribution he has made, to his new employer who continues to make deductions.

Holidays with pay The Building and Civil Engineering Holiday Scheme Management Ltd is responsible for the administration of the scheme. It is operated more or less on the principles of the National Insurance Acts: every operative has a holiday credit card. This card is the property of the employee, but they are kept by the employer, who is responsible for the card and the credit stamps fixed to it. In outline, the scheme requires the employer to purchase stamps from the management company and to fix a stamp on the employee's card for every complete week worked in his service. When the holiday period begins, the employer adds to the employee's pay packet the amount that he has in credit on the stamp cards and sends the card to the management company, who reimburse the employer with the full amount paid. The amounts paid by the employers under the scheme vary from time to time as the wage rates in the industry fluctuate, and the contributions are usually negotiated with every wage agreement between both sides of the industry.

As the employer's weekly contributions for holidays with pay become an emolument of labour, they must be taken into account in the analysis of unit rates.

In addition to the annual holiday entitlement, employees must be paid for the public holidays which fall outside their annual holidays. A similar credit stamp system used to be operated for this but it is now practice for the employer to cover himself in the same way as for inclement weather. The amount included in the 'all-in' rate is for the cost of the current annual holiday stamp plus an allowance for public holidays.

Severance or redundancy payments and sundry cost These costs will vary from firm to firm. Many building operatives do not stay with the same firm for the qualifying minimum period of two years. A builder might pay insurance premiums to cover his possible redundancy liabilities or include for his actual costs in a similar manner as guaranteed time allowances. Other factors such as fall in production during the period of notice, absenteeism, cost of insurances, etc., when an operative is absent later in the week after his cards were stamped on the Monday and other such instances combine to produce considerable costs under this heading.

Insurance for Employers' Liability, Third Party, and Injury to Persons and Property In the example given the cost of insuring for the above risks has been included in the all-in hourly rate. The premium will vary from firm to firm depending on the type of firm, type of work and claims history.

Trade supervision On most sites each trade will usually have a foreman or charge hand who is responsible for the supervision of that particular trade. The cost of this supervision will vary according to the number of operatives involved. In a small gang or trade the ganger or foreman will be working for the majority of his time whilst in a large group most of his time will be spent supervising and organizing.

Extra payments in accordance with the National Working Rule Agreement Provision is made in the National Working Rule Agreement for the Building Industry for certain additional payments to be made to operatives: these are divided into sections which are dealt with in the rules in great detail under the following headings:

1 Discomfort, inconvenience and risk (e.g. dirty money)
2 Continuous skill and responsibility (e.g. dumper drivers)
3 Intermittent responsibility (e.g. responsibility for the operation of a concrete mixer)

When analysing unit rates the estimator should add these 'plus rates' where applicable to the all-in hourly rate of every operative included in the analysis.

Apprentices Many building firms employ fully indentured craft apprentices who are paid wage rates proportionate to a craftsman's rate on the scale set out in the National Working Rule Agreement for the Building Industry. When estimating, it is impossible to decide whether a craftsman or an apprentice will execute the particular item of work being considered. Because of this difficulty, apprentices are disregarded when estimating and full craftsmen's rates are applied, on the basis that if an apprentice on half-pay executes the work, his output will be strictly proportionate to his wage rate: this cancels out the reduced rate of pay.

Apprentices are allowed two half-days per week with pay to attend a technical college and they receive full pay for holidays from their employer, as they are excluded from the holidays with pay scheme. Many firms include the time lost for technical education and holidays with pay for their apprentices in their overhead charges, but usually these costs are outweighed by the grants paid by the Construction Industry Training Board to firms who employ apprentices.

Incentives and bonus payments The National Working Rule Agreement for the Building Industry recommends that incentive schemes should be designed so that an operative of average ability and capacity should be given a reasonable opportunity of increasing his hourly rate earnings.

In practice there are several types of incentive bonus payments ranging from production-geared schemes to plain 'spot' bonus or combinations of these two extremes. The production incentive scheme means that an operative or group of operatives is given fixed 'targets'. The men are normally assured of the 'guaranteed minimum weekly earnings' and for work completed in excess of the target outputs they receive various fixed payments. The advantages include:

operatives have the opportunity to earn substantial extra money;
the builder can complete his contract sooner;

the builder should benefit by carrying out the work more cheaply than he envisaged reflected not only in the unit rates but in reduced overhead charges.

The disadvantages include:

frequent cases of faulty workmanship, in many cases only coming to light when the particular operatives have left the site;
the cost of rectifying such faults;
the difficulties and hidden costs of operating and supervising such schemes;
the problems connected with certain trades;
the disputes that can arise.

Whilst some firms can successfully claim to operate production-geared incentive bonus schemes, many other employers find it more practical to select their workforce and pay 'spot' bonus in the form of flat increases added to the minimum weekly rates. At the present time it is common to find many tradesmen particularly in urban areas being paid between £0.10 and £1.00 extra per hour and sometimes much more. Various combinations of 'spot' and incentive payments can be devised but this item has now become one of the most important elements of cost for estimators.

The question of why it is necessary to pay large bonus amounts to some operatives whilst there is high unemployment in the industry in the same localities poses an interesting problem, the answer to which lies outside the scope of this book.

While some estimators prefer to price the cost of bonus schemes in the preliminaries section of the bill of quantities, it is considered more usual and practical to include bonus payments in the build-up of the 'all-in' rate; this method will be used throughout this book.

Labour outputs Many builders' price books and text books on this subject list labour outputs in decimals of an hour, running to two or three places of decimals. It is felt by many that this is an entirely wrong approach to the subject and that the estimator should always think of labour in more tangible terms. Take as an example the fixing of architraves to doors: if the estimator assesses how long it will take to fix one set of architraves, and decides that the necessary time is about half an hour, this is far easier to appreciate than to think in terms of 0.20 hours per linear metre. Labour outputs whenever possible in this book will be based on specific operations which are easy to visualize, rather than expressed in decimals.

Some of the larger firms in the building industry have elaborate site-costing systems, which supply labour output data to the estimators. Such information can be of great help but great care has to be taken when using these figures as the output is likely to have been influenced by bonus payments.

Estimators should visit the sites as much as possible, constantly spot-checking labour outputs for items that they have previously priced. This is the only way to acquire the balanced judgement that goes to make a first-class estimator. In the building trade far too many estimators find themselves under constant pressure with the preparation of tenders and hardly ever set foot outside the office; this is obviously a very undesirable state of affairs.

Many trainee estimators, in their early days, think that the experienced estimator's view of labour outputs is extremely pessimistic; the trainees tend to think simply of the output that they could achieve if they did the work themselves. When told that a rate of 3¼ hours per cubic metre is reasonable for hand excavation, they realize that they could in fact do this in about two hours. But they forget that the operative concerned will be working day in and day out and that he will not be working with the enthusiasm of the young estimator; furthermore, he will not be digging one cubic metre only, but will most likely be digging continuously for eight hours. In addition to this, tea breaks, late starting and early finishing should be taken into account; it will then be seen that 3¼ hours per cubic metre is a reasonable figure.

Travelling expenses Travelling expenses are payable in accordance with the rules laid down in the Working Rule Agreement and in many cases allowances now become automatic. An average sum has been included in the 'all-in' rate build up.

Overtime The working of overtime up to one hour per day is now normal general practice except for midwinter and this means that the employer pays for these extra five hours per week at time and a quarter, that is, a total of 1¼ hours 'non-productive' cost per week. If no overtime is worked then the total cost of labour as previously illustrated would be divided by 40 to give a higher hourly rate. It follows that all parties benefit by the working of some overtime, always providing that the actual productivity does not drop at the end of the day.

2 Preliminaries

When preparing a tender for a contract based upon bills of quantities there are certain items of expense usually termed project overheads, that cannot be satisfactorily distributed among the unit rates; therefore some of these items are calculated as lump sums and are placed in the preliminaries bill. This chapter deals with the estimating of such items.

The most important items carrying the majority of cost in a typical building contract can be summarized under the following main headings:

1 General foreman
2 Insurances
3 Plant, tools, vehicles, etc.
4 Compounds and hoardings
5 Watching, lighting, security, etc.
6 Temporary services (water, electricity, telephone, etc.)
7 Temporary roads
8 Scaffolding
9 Temporary buildings
10 Guarantee bond or surety
11 Removal of rubbish
12 Special overtime
13 Special travelling expenses, lodging allowances, etc.
14 Firm-price adjustment

These items should also include the usual proportion of overheads and profit in the same way as any other unit rate.

In practice there is a wide difference between contractors in the way they show preliminaries costs. Some estimators prefer to include the majority of these costs in their unit rates throughout the bills of quantities, whilst others prefer to set out some of the labour emoluments, overheads and even profit in the preliminaries section.

Foreman, site staff, etc.

Some estimators prefer to include the foreman's or agent's salary in the firm's overhead charges; others prefer to include these costs directly in the tender. In this case, they are calculated by estimating the period of time during which the foreman will be present on the site, and applying the foreman's all-in weekly wage.

Some estimators prefer to use this method to deal with the costs of other staff who will be employed full-time on the site for long periods, such as timekeepers, storemen, costing clerks and even surveyors.

Insurance for fire risk, etc.

Clause 22 (A) of the JCT Form of Contract (1980 edition), when used for new buildings, states that the builder shall insure the building against fire risk, etc., to the value of the full contract sum, plus the percentage named in the appendix to the contract to cover professional fees. The premium charged for such insurance cover is usually included as a lump sum in the preliminaries section of the bill of quantities, and is calculated as follows:

Contract value	£500 000
Add percentage for fees, as appendix, 10%	50 000
	£550 000
Premium at, say, £0.15 per cent = £550 000 × 0.15% =	£825

Small plant

The erection of a building involves the use of numerous small items of plant, such as barrows, planks, picks, shovels and small hand tools. One of the simplest methods of including for such plant costs is to include in the preliminaries section of the bill of quantities a lump sum based upon the labour costs, as follows:

Total value of contract	£500 000
Labour cost included for firm's own work	£200 000
Small plant at, say, 0.75% of labour cost = £200 000 × 0.75% =	£1 500

An estimator will have access to records showing the actual costs of small plant over a period of time, which items are required, when they need replacing, etc. He can then include a calculation as shown above which will take into consideration a depreciation allowance and if necessary any unusual requirements of a particular contract.

Large plant

The costs of large plant can in some instances be included in the unit rates. For example excavating machinery in the rates for excavation, concrete mixers in the rates for concrete, etc., but there is a strong case for the majority of plant costs to be shown in the preliminaries section. Hoists, cranes, lorries and scaffolding will all probably be shared by at least several trades and it is difficult to apportion the cost between different items and sections.

In the event of variations, additional work, delays caused by the client or architect, etc., to a project, if all the plant has been priced in the preliminaries it is normally an easier task to 'prove' claims for loss and expense under clause 26 of the JCT Form of Contract. This factor can influence an estimator when he is preparing a tender and some contractors will include, for example, the cost of excavating and concreting plant in this section. The method of estimating for mechanical plant is shown in the next chapter, together with a discussion on the new plant items which appear in the majority of work sections of a bill of quantities as provided by SMM6.

Compounds and hoardings

It is often necessary to erect wire compounds around sites in isolated locations or just around the area where materials are stored. Such a cost would be made up of the supply and erection of concrete posts, fencing wire, post holes and bases, lock-up gates and removal costs. It is often necessary, in congested city centres and vandal-prone areas, to completely surround the site with a close-boarded hoarding with warning lights and provision for lock-up access points. Sometimes police permission has to be obtained to use the public footpath, and extensive provision must be made for temporary footwalks with guard rails and fan-shielding above. The cost of such an item would be made up of the supply and erection of timber posts, post holes and bases, exterior-quality plywood sheeting, lock-up entry requirements, maintenance of temporary lights, and provisions for removal. The salvage value of such items would have to be considered but in many instances would be negligible.

Watching, lighting and security

Certain types of building contracts require the attendance of a watchman during non-working hours, for part or occasionally the whole of the contract period. These costs are included in the preliminaries section of the bill of quantities as an estimated lump sum, and are calculated as follows:

Contract value	£500 000
Estimated contract period 50 weeks	
Estimated period of watching 30 weeks	
30 weeks at 9 shifts per week = 270 shifts at, say, £25.00 per shift ('all-in' rate) =	£6 750

(The remuneration of a watchman is at a rate per shift (day or night) calculated at one-fifth of the appropriate weekly rate for labourers.)

If warning lights are to be used an estimated sum will be added to cover this cost. The expenses of providing a hut, fuel, etc., would also have to be considered.

A more likely method of providing watching and security for a site would be to engage a specialist security firm who will provide a service appropriate to the contractor's requirements. They will visit the site at fixed minimum intervals outside working hours, provide a continuous manning by dogs or men or various other agreed combinations. Many building sites are plagued with damage by thieves and vandals and security is now very important and extremely expensive. A traditional watchman is not the answer. Legislation has now been introduced which forbids guard dogs to be left loose and unattended on sites and this can only mean an even greater rise in the costs under this heading.

Works water supply

Very few building sites can operate without a water supply. It is very difficult to include the cost of water in each unit rate; it is far easier to include it as an estimated lump sum in the preliminaries section of the bill of quantities. On large contracts, the water supply is metered and charged by the local authority on the basis of the amount used. On smaller contracts, a lump-sum charge could be made by the water authority after viewing the drawings. For estimating purposes, water costs can be calculated as a percentage of the contract sum, using previous contracts as a basis. A fairly representative figure would be about ⅛ of 1 per cent, and to this figure must be added the cost of temporary plumbing, as follows:

Total value of contract £500 000

Estimated cost of water, ⅛%	£625.00
Excavation of trench for pipe, 20 metres at £5.00 per metre	100.00
28 mm copper service pipe, 20 metres at £5.00 per metre	100.00
Local authorities charge for connection to main, say	50.00
Standpipe, tap and box with lock, say	60.00
Clearance and making good, say	35.00
Total	£970.00

Temporary electricity supply

The cost of this item depends on the type of contract, the duration over different seasons of the year, the type and extent of plant, equipment and lighting requiring electrical energy, etc. The connection charge is usually free unless a supply is not available in the near locality.

The cost therefore would be based on the number of units consumed and is paid quarterly. An estimator would have to use his experience and judgement on the likely expense involved. Temporary wiring, fittings and low voltage equipment would also have to be considered.

Temporary telephone

This item has a similar basis to electricity, but there is always a connection charge of at least £50, quarterly rentals from £15 and the charges for calls made. An estimator would have to consider who would be using the telephone and whether the calls were mainly local or long-distance.

Temporary roads

If, due to site conditions, it is necessary to construct a temporary road for access to the site, the cost will be calculated and inserted in the preliminaries. The cost of such an item would be made up of the use and waste of sleepers or hardcore, etc., labour laying and removing, transporting debris from the site and making good after removal.

Scaffolding

Methods of pricing scaffolding are given in Chapter 3.

Site offices and amenity buildings

The cost of these temporary buildings can be estimated on a rental basis for the period that they will be required on the site. This sum is included in the preliminaries section of the bill of quantities and is estimated as follows:

Total value of contract £500 000
Estimated contract period 50 weeks
Maintenance period 6 months or 26 weeks

2 site offices 3 m × 4 m at £5.00 per week	£10.00
1 hut for men 6 m × 6 m at £10.00 per week	10.00
1 hut for materials 9 m × 6 m at £15 per week	15.00
1 office for clerk of works 3 m × 4 m at £6.00 per week	6.00
2 chemical latrines at £3.00 per week	6.00
Cost per week	£47.00

76 weeks (50 + 26 maintenance, if required) at £47.00	£3572
Add rates on temporary buildings, say 10%	357
Transport to and from site – 5 tonne lorry, two days at £75.00	150
Four joiners, erecting, maintaining and dismantling, say a total of one week = 4 × 40 = 160 hours at £5.00 per hour	800
Attendant labourer 76 weeks at 2 hours (average) per day = 76 × 2 × 5 = 760 hours at £4.00 per hour	3040
Furniture and fittings required 76 weeks at, say £10.00 per week	760
Total	£8679

It may be that the rental agreement includes transport, erection, maintenance and removal from site in which case some of the above calculations would not be shown but would be reflected in higher rent charges.

If the contractor owns the huts, furniture, fittings, etc., he would have to calculate a depreciation cost which would replace the hire rates shown in the example. If the client requires a particular temporary building including fittings the contractor would have to consider this cost very carefully.

Guarantee bond

It is usual on many contracts for the client to require a guarantee bond or other surety. A contractor normally pays a single insurance premium for this and it can be a very expensive item depending on the financial status and experience of the contractor. A cost in the region of £5000 for a bond on a £500 000 contract is currently possible.

Removal of rubbish

This would include periodic tidying up as well as at the completion stage. An example of this would be as follows:

Contract period: 50 weeks	
Allow labourer 2 hours (average) per week for tidying up and loading rubbish = 50 weeks × 2 hours = 100 hours at £4.00	£400
At completion allow 2 men for 1 day = 16 hours at £4.00	64
Total	£464

This calculation could be supplemented by the expenses of dumpers, lorries, etc., if they were not separately priced in the preliminaries. Charges for tipping fees must also be considered.

Overtime

The National Working Agreement lays down a set of rules for calculating the payment of overtime which include the following provisions: 'Time and a quarter for the first hour, time and a half for the next two hours and double time thereafter until starting time next morning. Time worked between starting time and 4 p.m. on Saturday is paid at time and a half and afterwards at double time until starting time on Monday morning.'

An allowance for overtime worked as normal practice may be included in the all-in hourly rate shown in the previous chapter, if this is not done the cost of unproductive overtime can be calculated and included in the project overheads according to the anticipated overtime which will not be specifically reimbursed by the client.

When dealing with estimates for work which is to be carried out after normal working hours and at week-ends, such as shop and bank alterations, the unproductive overtime can be included by using an enhanced all-in rate.

Travelling expenses

An average normal allowance for travelling expenses has been included in the all-in rate as shown in the previous chapter. An estimator will have to consider each particular contract and refer to the Working Rule Agreement.

Under the National Working Rule Agreement for the Building Industry an operative shall use either public transport or transport provided by the employer, at the employer's option. Many of the larger firms of contractors own or hire bus transport for their operatives, and a sum must be included in the preliminaries section of the bills of quantities to cover the cost. It can be estimated simply by calculating the number of daily outward and inward journeys, with the seating capacity of each vehicle, and the number of vehicles required.

Some of the smaller and medium-sized firms in the industry transport workpeople in lorries equipped with suitable all-weather equipment. If such vehicles are used extensively for the transport of men, the cost is charged for the inward and outward journeys, that is, two a day, but if the lorries can be utilized on the site during the day, the cost can be included elsewhere.

Lodging allowances

Sometimes a site is situated so far away from the firm's headquarters that the operatives have to find lodgings. In such cases the National Working Rule Agreement for the Building Industry stipulates a minimum sum per night that must be paid to each operative. In practice, most firms find that they have to pay more than this amount as an incentive for operatives to stay with the firm when sent to contracts away from home. An estimate of this cost can be calculated by multiplying the anticipated number of man weeks by a calculated weekly lodging allowance. An additional sum will also be added to cover the fares for travelling home periodically. On some contracts it may even be necessary to establish a temporary 'camp' involving the provision of sleeping, eating, and recreational facilities.

Firm-price contracts

Before the Second World War practically all building contracts were arranged on a firm-price basis. Due to the inflation of the war years and the post-war period, it became necessary to incorporate fluctuations clauses in contracts to reimburse builders for any increases in the cost of labour and materials that took place during the contract period.

In 1958 the Ministry of Works stated that they intended in future to let all contracts of £100 000 and below on a firm-price basis. Other government departments, local authorities and private employers quickly followed suit and for some years the majority of contracts were let on this basis.

However, the extensive legislation of the mid 1960s which included heavy increases in National Insurance employers' contributions to the inception of Selective Employment Tax brought about the re-introduction of the fluctuation clause in many building contracts. During the return of firm-price tendering an effort was made in wages negotiation to agree wage increases phased over a three-year period. This method of wage negotiation was a great help to builders' estimators in their work, as it meant that they were aware of the proposed wage increases and the dates on which they were due to take effect. When preparing firm-price tenders, estimators included approximate sums in their tenders to cover the cost of wage and material increases.

In the period 1973–75, as inflation soared beyond the 25 per cent per annum mark, many contractors lost heavily on firm-price contracts especially in respect of increases in the cost of materials. During that period it became the normal rule that all projects of over twelve months duration were to be on a fluctuating-price basis. Contractors are currently reluctant to offer firm-price tenders for any work even of six months duration because of the uncertainty with future wage and material prices.

At the beginning of 1981 an estimator submitting a firm-price tender would attempt to include approximate sums to cover future labour increases as follows:

Date of tender 2 January 1981
Date for commencement of work 1 March 1981
Present craftsman's 'all-in' rate, say, £3.50 per hour
Contract period 12 months
Anticipated rise in wage rates and other labour emoluments from 1 July 1981 resulting in a revised 'all-in' rate of £4.00 per hour.

1 March to 1 July is 4 months and the contract period is 12 months; therefore there will be two-thirds of the contract period to run on 1 July. Assuming the main contractor's 'all-in' labour cost is £200 000, the estimated labour expenditure after July 1st:

$$= \frac{2}{3} \times \text{£}200\,000 = \frac{\text{£}133\,333}{\text{£}3.50} = 38\,095 \text{ hours at £}0.50 \text{ per hour increase}$$

$$= \underline{\text{£}19\,047.50}$$

In practice there would probably be different rates for alterations of different parts of the 'all-in' rate but an estimator must take an overall view of the likely situation based on his views for the future. In July 1975 most estimators would have been thinking of a possible 30 per cent increase in the next twelve months, while in January 1981 a figure of between 10 and 20 per cent was probable for the next year. This suggested method is one of many possible alternatives but the basis must always be an anticipated forecast on the information available with varying arithmetical applications.

As the sum included to cover wage increases is only approximate, no attempt is made to calculate the sum in two parts – one for craftsmen, and one for labourers – even though the wage increases may be known exactly. An attempt to allocate the sum between craftsmen and labourers would be pointless and too theoretical: the sum is therefore calculated on the higher increase only.

When preparing tenders for firm-price contracts, it is difficult to calculate a sum to cover the possible rises in materials prices over the contract period. The only practical way to overcome this difficulty is to ask materials suppliers to quote prices for materials which will remain static for the contract period.

In many cases manufacturers will only quote for current costs and an estimator must take the price of each material in turn, the national inflation position, the likely date in the future when he will need to purchase the materials and by considering how inflation will affect this particular item, add a lump sum or percentage to the quoted price.

From the details shown it is quite obvious why there is a widespread reluctance to give any firm-price commitment beyond a few months. Some estimators prefer to allocate this fluctuation risk to the items that will be particularly affected in the later stages of a project.

When preparing tenders for contracts with fluctuations clauses, the estimator's work is eased considerably by the fact that he bases his unit rates on the wage rates and materials price current at the date of tender, with the knowledge that all subsequent wages fluctuations and changes in materials prices listed in the basic price list will be later reimbursed net to the firm. To make tendering easier in such cases, and to overcome the difficulties caused by last-minute changes in wage rates and prices, an operative date of tender which is not less than ten days before the date due for the receipt of tenders is used.

The current position with the JCT Form of Contract is that clause 40 was introduced late in 1975 and provides a formula price adjustment for fluctuations in costs of labour, materials and overheads based on indices. In theory a contractor can tender confidently for any project on current costs and be satisfied that he will be reimbursed for all increases including his indirect charges. This clause is an optional alternative to the traditional system of claiming wage and material increases by means of verified weekly site records and invoices and comparing them with a basic price list submitted with the tender (clause 39). The net differences would be subject to a percentage addition inserted against clause 39.8 which is meant to cover the contractor's administration costs involved with the recovery of the fluctuations. The actual percentage included for 39.8 is usually decided by the client and the tendering contractor must decide whether or not this amount will be sufficient to cover his fluctuation risk not only in respect of administration overheads but also for minor materials and other overheads and profit which he will not be reimbursed for elsewhere in accordance with clause 39. Although clause 39 has been modified to allow a slightly wider recovery of fluctuations, this method still does not reimburse a contractor for large parts at the risk of inflation, in particular, to items of plant and the majority of site overheads.

In conclusion a contractor when estimating the costs of a fluctuating contract can never be satisfied that all the fluctuation risk is covered whether clause 40 is used or not. There must always be a proportion of his tender sum which is at the risk of inflation. He must attempt to calculate this amount, consider the likely inflation rate and include a sum which will be sufficient.

Value Added Tax (VAT)

The introduction of this tax in 1973 has obviously affected the building industry. At the present time new construction work has a zero rating which means that there are no direct extra costs involved. Although suppliers' and sub-contractors' accounts carry VAT surcharges to the main contractor, this is claimed back from the Customs and Excise Authority and not passed on to the client. There are several minor exceptions to this rule and a client would have to pay the VAT element on certain odd items, e.g. a fixed electric fire. Unfortunately the administration cost, etc., of paying and collecting VAT and refunds is considerable and could possibly make an additional cost to building work in the region of 1 per cent of the contract amount. This charge would normally be included in the allowance for overheads.

On the other hand, building work of a maintenance or repair nature carries a positive VAT rating and unless the client is 'Vat-able' will be a direct extra charge. The exact dividing line between 'improvement' and 'maintenance' has caused great difficulty amongst builders and quantity surveyors.

Other items

It is possible for estimators to include sums in the preliminaries section for other items, notably setting out, defects liability, special access to the site, and of course the phasing of work if required.

3 Mechanical plant

Costing

There are several methods of costing mechanical plant in use today, some of which are extremely complicated and mainly theoretical, taking into account grants and tax allowances for capital investment and 'second-hand' values. The methods shown later in this book are very simple, practical and effective, and are used by many plant-hire firms to calculate their hire charges.

When a building firm maintains a separate plant section it is the usual practice to debit the contracts with a charge for plant supplied, on an hourly basis, irrespective of whether the plant is working or not. This method gives every incentive for the site agent to use the plant as quickly as possible, and return it promptly to the depot. Such firms often have two hourly rates for plant: one for internal use, which does not carry any profit, and another for external hire, which includes a profit charge. There are, however, some firms which run their plant department on a separate basis and which charge the same hourly rate for internal and external use.

Hire of plant

Many building contractors hire all their plant requirements from specialist plant hirers. The advantage of this method is that plant can be hired for the minimum period necessary, and the building firm does not have the difficulty of having to find work for plant standing idle. In addition, it does not have to engage skilled fitters and special staff who thoroughly understand the business of plant operation and maintenance. One disadvantage of hire as against ownership is the fact that it is often more costly in the long run to hire than to own.

Ownership of plant

There are two main factors which normally decide whether a firm will hire or own plant; these are the finance available and the amount of work calling for machinery. Idle plant means financial loss; therefore an owner of mechanical plant must have sufficient work for the plant, either through the firm's own contracts, or through public hire business. Building and civil engineering plant is extremely expensive, but finance need not play such a great part these days, due to the growth of hire-purchase facilities. A thriving plant-hire business can afford to pay off hire-purchase agreements, and still make a profit out of the hire rates.

Grants and tax allowances affect the decision of whether or not to purchase plant. These tax allowances vary from time to time depending upon the economic position of the country and also the locality of the contractor's organization. An overall nominal amount is shown in the examples to cover net interest, net taxes, insurance, etc.

Depreciation

Actual depreciation figures can be obtained from past records, when a firm has successfully run a plant-hire department for a number of years. Depreciation can be calculated by dividing the purchase price of the machine by the number of hours of working life. Below is a list showing the average working life in hours of several items of plant.

	hours		*hours*
Mechanical barrows	5000	Excavators	10 000
Paint-spraying equipment	6000	Tractors	10 000
Dumpers	8000	Lorries	10 000
Concrete mixers	8000	Compressors	10 000
Hoists	8000	Rollers	15 000
Pumps	8000	Cranes	15 000

Interest, taxes and insurance

When calculating hourly rates for plant, the interest on the money invested in the plant should be taken into account. This is the interest that the money would have earned had it been invested in the gilt-edged stock market. Several items of plant are taxed in the form of Road Fund Tax, and all plant should be insured for the full replacement value to cover possible loss due to damage, or even theft. It will be found that the total of these three items amounts to approximately 10 per cent of the hourly depreciation charge for practically all items of plant.

Repairs and maintenance

The upkeep of mechanical plant is very expensive; rough conditions on site cause extensive damage, and spare parts for such complex machines are very costly. Accurate figures for the amounts expended in plant maintenance can be calculated when firms have been operating plant for a number of years; these figures should always be used. Below is a list of the average maintenance costs of various items of plant, expressed as a percentage of the hourly depreciation charges.

	%		%
Rollers	10	Dumpers	25
Paint-spraying equipment	10	Hoists	25
Pumps	10	Compressors	25
Lorries	15	Excavators	33⅓
Cranes	20	Tractors	33⅓
Concrete mixers	20		

Tyres

Wear and tear on tyres for many types of wheeled mechanical plant can be very costly. Naturally, factors such as maintenance, speeds, surface conditions, wheel positioning and loads affect tyre life, but on average the life of a tyre on mechanical plant is between 3000 and 4500 hours. When calculating hourly rates, tyre maintenance should be taken into account separately from mechanical maintenance. There is a large variation in prices for the renewal of tyres; a set of tyres for a small dumper costs about £200 for a six-wheel lorry about £1200, for a wheeled excavator about £1000, and possibly over £10 000 for a large scraper. These costs can be reduced a little if remould tyres are used.

Fuel and lubricants

Site conditions affect the fuel consumption of plant, therefore only average consumption figures can be used. Tables of average fuel consumption can be obtained from various sources, but once again accurate records compiled when actually operating plant should always be used in preference. It should be noted that plant owners can obtain diesel fuel oil exempt from duty if they store it in an approved manner and keep proper records.

Plant labour costs

The drivers of mobile plant are usually permanently assigned to their machines and attend them wherever the machines go. They are paid more than the basic rate for labourers, and in addition they often earn considerable bonuses and receive travelling and lodging allowances. The operators of stationary plant are often labourers employed on general work who are assigned to operate the plant when the need arises.

For certain types of excavation, excavators often require a labourer to direct the machine operator and to trim off small amounts of excavation by hand; these labourers are known as banksmen. When calculating hourly costs of mobile plant, the operative's wages are usually included in the rate, but with stationary plant the cost is calculated exclusive of the attending operative, whose wages are charged in with the labour gang (for example, concrete mixers) engaged on the operation. The National Working Rule

Agreement for the Building Industry sets out a comprehensive list of all the plus rates paid to operatives of mechanical plant, over and above the labourer's rate.

Transport of plant

It is very difficult to include the costs of transport of plant to and from site in the hourly rate. These costs increase the hourly rate appreciably when plant is required on site for one week only, but represent only a small fraction of the hourly rate when plant is required on site for one year.

The most satisfactory method is to include the plant transport costs as a lump sum in the preliminaries bill. Most mobile plant and large static plant has to be transported on a low loader; the daily charge for a low loader varies between £80 and £150 per day according to weight capacity. The costs of offloading, setting up, and moving plant about the site can be quite heavy, particularly for certain types of lifting plant that require tracks. Static hoists require several man hours to erect and tie into the scaffolding, and the erection and dismantling of the large monotype tower crane can cost as much as £15 000.

Costs of owning and operating a 5/3½ open drum petrol concrete mixer with an output of 0.10 m³ per mix

Owning costs

Cost £1200		
Depreciation on average life condition of 8000 hours $= \frac{£1200}{8000}$		£0.150
Interest, taxes and insurance, 10%		0.015
Repairs and maintenance, 20%		0.030
Hourly owning cost		£0.195

Operating costs

Fuel per 8-hour day, 7 litres at 25p	£1.750	
Lubricating oil, 3 litres per week at 80p $= \frac{£2.40}{5 \text{ days}} =$ per day	0.480	
Fuel and oil per day	£2.230	
Fuel and oil per hour $= \frac{£2.230}{8}$		0.279
Hourly owning and operating costs		£0.474

Costs of owning and operating a 10/7 diesel concrete mixer with an output of 0.20m^3 per mix with scraper shovel, batch weigher and aggregate feed apron

Owning costs

Cost £3400

Depreciation on average life condition of 8000 hours

$= \frac{£3400}{8000}$	£0.426
Interest, taxes and insurance, 10%	0.043
Repairs and maintenance, 20%	0.085
Hourly owning costs	£0.554

Operating costs

Fuel per 8-hour day, 7 litres of diesel oil at 15p	£1.050	
Lubricating oil, 4 litres per week at 80p		
$= \frac{£3.20}{5 \text{ days}} =$ per day	0.640	
Fuel and oil per day	£1.690	
Fuel and oil per hour $= \frac{£1.690}{8}$		0.211
Hourly owning and operating costs		£0.765

Costs of owning and operating a one-cubic-metre-capacity diesel dumper

Owning costs

Cost £2300

Depreciation on average life condition of 8000 hours

$= \frac{£2300}{8000}$	£0.288
Interest, taxes and insurance, 10%	0.029
Repairs and maintenance, 25%	0.072
Hourly owning costs *c/fwd*	£0.389

Operating costs

Fuel per 8-hour day, 5 litres of diesel oil at 15p	£0.750
Lubricating oil, 4 litres per week at 80p	
$= \frac{£3.20}{5 \text{ days}} =$ per day	0.640
Fuel and oil per day	£1.390

	b/fwd	£0.389
Fuel and oil per hour = $\frac{£1.390}{8}$		0.174
Tyres, two sets at £200 per set = £400		
Cost per hour = $\frac{£400}{8000}$		0.050
		0.613
Operator's wages, 'all-in' labourer's rate of £4.00 per hour, plus allowances in accordance with the National Working Rule Agreement (NWRA), say		4.250*
Hourly owning and operating costs		£4.863

Costs of owning and operating a JCB3 ¼ cubic metre diesel tractor excavator

Owning costs

Cost £24 000		
Depreciation on average life condition of 10 000 hours = $\frac{£24\ 000}{10\ 000}$		£2.400
Interest, taxes and insurance, 10%		0.240
Repairs and maintenance, 33⅓%		0.800
Hourly owning costs		£3.440

Operating costs

Fuel per 8-hour day, 16 litres of diesel oil at 15p	£2.400	
Lubricating oil, 5 litres per week at 80p = $\frac{£4.00}{5\text{ days}}$ = per day	0.800	
Fuel and oil per day	£3.200	
Fuel and oil per hour = $\frac{£3.200}{8}$ = per day		0.400
Tyres, two sets at £1000 per set = £2000		
Cost per hour = $\frac{£2000}{10\ 000}$ =		0.200
Operator's wages, 'all-in' labourer's rate of £4.00 per hour, plus allowances in accordance with the NWRA, say		4.250*
Hourly owning and operating costs		£8.290

*See plant labour costs, page 39.

Costs of owning and operating a D6B Caterpillar crawler tractor hauling an 8/10 scraper

Owning costs

Tractor	£50 000
Cable control	6 000
Scraper	14 000
	£70 000

Depreciation on average life condition of 10 000 hours $= \frac{£70\,000}{10\,000}$	£7.000
Interest, taxes and insurance, 10%	0.700
Repairs and maintenance, 33⅓%	2.333
Hourly owning cost	£10.033

Operating costs

Fuel per 8-hour day, 80 litres of diesel oil at 15p	£12.000	
Lubricating oil, 10 litres per week at 80p $= \frac{£8.00}{5 \text{ days}} =$ per day	1.600	
Grease per day = 3 kg at £1.00 per kg	3.000	
Fuel and oil per day	£16.600	
Fuel and oil per hour $= \frac{£16.600}{8} =$		2.075
Tyres for scraper, one set at £10 000 Cost per hour $= \frac{£10\,000}{10\,000} =$		1.000
Operator's wages, 'all-in' labourer's rate of £4.00 per hour plus allowances in accordance with the NWRA, say		4.250
Hourly owning and operating costs		£17.358

Costs of owning and operating a 3.5 m³ per minute diesel portable air compressor, equipped with medium-duty concrete breakers and 16 m of unarmoured hose

Owning costs

Cost complete with tools £7000

Depreciation on average life condition of 10 000 hours

$= \frac{£7000}{10\ 000}$ £0.700

Interest, taxes and insurance, 10% 0.070

Repairs and maintenance, 25% 0.175

Hourly owning cost £0.945

Operating costs

Fuel per 8-hour day, 22 litres of diesel oil at 15p £3.300

Lubricating oil, 2 litres per week at 80p

$= \frac{£1.60}{5 \text{ days}} =$ per day 0.320

Fuel and oil per day £3.620

Fuel and oil per hour $= \frac{£3.620}{8}$ 0.453

Operator's wages, 'all-in' labourer's rate of £4.00 per hour plus allowances in accordance with the NWRA, say 4.250

Hourly owning and operating costs £5.648

Costs of owning and operating a 7-tonne (5 cubic metre capacity) petrol motor lorry for medium-distance contract work

Owning costs

Cost £12 000

Depreciation on average life condition of 10 000 hours

$= \frac{£12\ 000}{10\ 000}$ £1.200

Interest, taxes and insurance, 15% 0.180

Annual Road Fund Tax for 6 years at, say £500 per annum

Cost per hour $= \frac{£3000}{10\ 000}$ 0.300

Repairs and maintenance, 15% 0.180

Hourly owning costs *c/fwd* £1.860

Operating cost

Fuel per 8-hour day, 44 litres of petrol at 30p £13.200

Lubricating oil, 6 litres per week at

80p $= \frac{£4.80}{5 \text{ days}} =$ per day 0.960

Fuel and oil per day £14.160

Hourly owning costs *b/fwd*	£1.860
Fuel and oil per hour $= \dfrac{£14.160}{8}$	1.770
Tyres, four sets of six tyres at £1200 per set = £4800	
Cost per hour $= \dfrac{£4800}{10\ 000}$	0.480
Driver's wages, 'all-in' labourer's rate of £4.00 per hour plus allowances in accordance with the NWRA, say,	4.250
Hourly owning and operating costs	£8.360

Scaffolding

In estimating for building work, there are two methods of assessing scaffolding costs to be included in the tender. One method is to calculate the hire, erection and dismantling costs for the estimated period of use and to include this sum in the preliminaries section of the bill of quantities.

The second method is to include the scaffolding costs with the unit rate for each square metre of brickwork. This method is comparatively simple when the scaffolding required encloses the external brick walls of a building of traditional design (e.g. a brick-built house). However, if the external walls of a building are not entirely comprised of brickwork, the brickwork measurements will not reflect the true scaffolding costs.

Tubular steel scaffolding required for 100 square metres area on the face of the building (estimated period of use: 3 months)

Hire charges per month	
190 metres steel tubing at 8p per metre per month	£15.20
14 metres steel putlogs at 7p per metre per month	0.98
50 couplings at 5p each per month	2.50
12 base plates at 10p each per month	1.20
90 metres of scaffold boards required for one lift only at 20p per metre	18.00
Hire charge for one month	£37.88

Hire charge for a 3-month period £37.88 × 3 =	£113.640
Add 5% for breakages and loss	5.682
A scaffolder will erect about 8 m² of tubular steel scaffold per hour, at the 'all-in' craftsmen's rate, in accordance with NWRA. 12½ hours at £5.00 per hour	62.500
A scaffolder will dismantle about 25 m² of tubular steel scaffold per hour. 4 hours at £5.00 per hour	20.000
Cost of erecting, dismantling and hiring a tubular steel scaffold for an area of 100 m² for three months	£201.822

Total area of scaffolding for the contract assumed to be
1000 m² therefore total cost =

$$£201.822 \times \frac{1000}{100} = £2018.22$$

Generally, however, scaffolding will be carried out by specialist sub-contractors who will offer a tender figure (or rates) for the work involved, based on the main contractor's programme and the drawings.

Care must also be taken to make suitable allowances for all internal scaffolding of a minor nature, for example portable trestles for partitions, plastering, painting, etc.

There are many other possible items of mechanical plant which need to be considered by an estimator including all types of hoists and cranes, pumps, vibrators, rollers, generators, power tools, floodlighting, etc.

This chapter illustrates how mechanical plant can be costed, but as previously mentioned, contractors will insert the costs of plant, fuel operators, insurances, etc., against differing items in the bills of quantities according to their particular viewpoints.

In conclusion there are three main ways of pricing the plant costs in bills of quantities:

1 To calculate the costs for some or all items of plant and include such sums against SMM6 clause B.13.1a.
2 To include some or all of the costs of plant against appropriate 'quantity-related' items in the bills, such as trench excavation, hardcore filling, concrete in foundations, etc.
3 To include some or all of the costs against the particular plant items provided at the beginning of most sections of a bill of quantities (e.g. SMM6 clauses C.2, D.4, E.3, F.2, etc.).

Example based on SMM6 clause D.4 with a contract requiring a JCB3 type excavator

Clause D.4.1

	£
Low loader hire charge for bringing plant to site, minimum charge, 4 hours at £12.00	48.00
Low loader hire charge for removal of plant from site, minimum charge, 4 hours at £12.00	48.00
Total cost	£96.00

Clause D.4.2

Assume that the excavator has a working hire charge of £9.00 per hour either from a subsidiary company or specialist plant hire firm including fuel and operator. (The fuel and operator can be estimated separately if required.)

Allow an operational use of 100 hours at £9.00	£900.00

These two items are examples of 'method-related' costs which have been introduced in SMM6 based on the philosophy adopted in the 1976 Civil Engineering Standard Method of Measurement. The first item is a 'fixed' cost and the second is a 'time-related' cost. The estimator when pricing the bills of quantities would decide upon the plant required for a particular trade and how long it will be needed on site. The settlement of the extra costs of plant due to variations or delays caused by the client will then be more readily achieved.

The concept is good in theory but it may well be that estimators will not accept this method of pricing through reluctance to expose these details of a tender or in view of the complications of trying to split the costs of, say, an excavator between the plant items of various sections of the bills, for example, substructures, drainage, pavings, external services, etc. The precast concrete example quoted in the introduction to SMM6 Practice Manual is a more likely possibility for pricing plant items. Perhaps SMM7 will consider the pricing of plant in a more realistic manner.

4 Excavation and earthwork

Machine excavation

Excavate to reduce levels, maximum depth not exceeding 2 m. Per cubic metre (using a ¼ cubic metre excavator, loading direct into lorries for disposal)

The output of an excavator loading direct into lorries is dependent upon the disposal facilities. An excavator working non-stop for an hour, with ample capacity for spoil, will give a very good output. However, an excavator working with only one lorry, which has to haul the spoil over a long distance, will have a low output because of the standing time. To employ an excavator to its full capacity, ample lorry capacity must be available. A ¼ m^3 excavator, excavating to reduce levels in ordinary ground under normal conditions, should give an average output of between 7 and 9 m^3 per hour depending on the skill of the operator, angle of swing, depth, etc.

	£
Cost per hour of a ¼ m^3 excavator, with operator as previously analysed	8.29
At an output of 8 m^3 per hour,	

$$\text{cost per m}^3 = \frac{£8.29}{8} = \underline{£1.04}$$

Unit rate = £1.04 per m^3 (labour nil, plant £1.04)

Excavate basement, starting at ground level, maximum depth not exceeding 2 m. Per cubic metre (using a ¼ cubic metre excavator, loading direct into lorries for disposal)

Outputs for basement excavation are a little higher than those for shallow, reduced-level excavation. This is due to the fact that, when working to reduced levels, the machine is covering a larger area to a shallower depth. With basement excavation, the machine will win 90 per cent of the spoil with large loads and it is only the last 10 per cent that requires extra care when working to a fixed level. A ¼ m^3 excavator, excavating to basements not exceeding 2 m deep in ordinary ground under average conditions, should give an output of about 9 m^3 per hour.

	£
Cost per hour of a ¼ m³ excavator with operator, as previously analysed	8.29
A banksman will be required for this type of excavation, so labourer 1 hour at £4.00	4.00
Total cost per hour	£12.29

At an output of 9 m³ per hour,

$$\text{cost per m}^3 = \frac{£12.29}{9} = \underline{£1.366}$$

Unit rate = £1.37 per m³ (labour £0.45, plant £0.92)

Excavate basement, starting at ground level, maximum depth not exceeding 4 m. Per cubic metre (using a ¼ cubic metre excavator, as before)

Owing to the extra depth, the output of the machine will fall to about 7 m³ per hour.

	£
Cost per hour of a ¼ m³ excavator with operator and banksman, as previously analysed	12.29

At an output of 7 m³ per hour,

$$\text{cost per m}^3 = \frac{£12.29}{7} = \underline{£1.756}$$

Unit rate = £1.76 per m³ (labour £0.57, plant £1.19)

Excavate foundation trench, starting at ground level, maximum depth not exceeding 1 m. Per cubic metre (using a ¼ cubic metre excavator as before)

As the excavator is working in a confined area and to an accurate width, the output will fall to about 6 m³ per hour.

	£
Cost per hour of a ¼ m³ excavator with operator and banksman, as previously analysed	12.29

At an output of 6 m³ per hour,

$$\text{cost per m}^3 = \frac{£12.29}{6} = \underline{£2.048}$$

Unit rate = £2.05 per m³ (labour £0.67, plant £1.38)

Excavate foundation trench, starting at ground level, maximum depth not exceeding 2 m. Per cubic metre (using ¼ cubic metre excavator as before)

Owing to the extra depth, the output of the machine will fall to about 5 m³ per hour.

	£
Cost per hour of a ¼ m^3 excavator with operator and banksman, as previously analysed	12.29

At an output of 5 m^3 per hour,

$$\text{cost per m}^3 = \frac{\pounds 12.29}{5} = \underline{\pounds 2.458}$$

Unit rate = £2.46 per m^3 (labour £0.80, plant £1.66)

All the mechanical excavation items analysed so far have been based upon excavation in ordinary ground. If different strata are encountered, the ordinary ground output will be reduced approximately to the following percentages:

Heavy clay – 75% of ordinary ground output
Soft chalk – 50% of ordinary ground output
Hard rock – 20% of ordinary ground output

Remove excavated material a distance not exceeding 100 metres and deposit in spoil heap. Per cubic metre (using a 1 cubic metre dumper)

Working under average conditions, a 1 m^3 dumper should remove about 5 m^3 of spoil 100 metres in one hour. As spoil bulks approximately 25%, these 5 m^3 of loose spoil represent 4 m^3 in the ground, and as all items of excavation and subsequent disposal in bills of quantities are composed of the actual quantities before excavating, the output is 4 m^3 per hour.

	£
Cost per hour of a 1 m^3 dumper, with operator as previously analysed	4.863

At an output of 4 m^3 per hour,

$$\text{cost per m}^3 = \frac{\pounds 4.863}{4} = \underline{\pounds 1.216}$$

Unit rate = £1.22 per m^3 (labour nil, plant £1.22)

Remove surplus excavated material from site. Per cubic metre (using a 5 cubic metre lorry hauling a distance of 3 kilometres)

Lorry load = 5 m^3 of bulked spoil = say 4 m^3 in the ground.

Standing time loading		20 minutes
Travelling to tip		15 minutes
Tipping		10 minutes
Return from tip		15 minutes
	Total time	60 minutes

£

Cost per hour of a 5 m³ lorry, with driver as previously analysed — 8.36

At an output of 4 m³ per hour,

$$\text{cost per m}^3 = \frac{£8.36}{4} = \underline{£2.09}$$

Unit rate = £2.09 per m³ (labour nil, plant £2.09)

Note: Should any tipping charges be incurred they would have to be added to the above cost. If the excavation was not loaded into the lorries initially by the excavator it would be necessary to include in this item for the cost of digging in spoil heaps and loading.

Excavate over area of site average 150 mm deep to remove vegetable soil. Per square metre (using a ¼ cubic metre excavator)

The output of a ¼ m³ excavator, stripping the area of the site to shallow depths, will be about 6 m³ per hour. Cost per hour of ¼ m³ excavator, and operator as previously analysed.

At an output of 6 m³ per hour,

$$\text{cost per m}^3 = \frac{£12.29}{6} = \underline{£2.048}$$

$$\text{Cost for 1m}^2 \text{ 150 mm thick} = £2.048 \times \frac{150}{1000} = \underline{£0.307}$$

Unit rate = £0.31 per m² (labour nil, plant £0.31)

Hand excavation

Excavate to reduce levels, maximum depth not exceeding 2 m. Per cubic metre (using hand labour, loading into barrows for disposal)

A labourer will excavate 1 m³ of ordinary ground in about 2½ hours.

£

Cost per m³ = 2½ hours at £4.00 — 10.00

Unit rate = £10.00 per m³ (labour £10.00)

Excavate over area of site 150 mm deep to remove vegetable soil, per square metre (using hand labour)

A labourer will excavate 1 m³ in about 2 hours, using single-spit digging and throwing into barrows.

£

Cost of excavating per m³ = 2 hours at £4.00 — 8.00

Cost for 1m² 150 mm thick = $£8.00 \times \frac{150}{1000} = £1.20$

Unit rate = 1.20 m² (labour £1.20)

Excavate foundation trench, starting at ground level, maximum depth not exceeding 2 m. Per cubic metre (using hand labour)

A labourer will excavate 1 m³ in ordinary ground in about 3¼ hours; this includes throwing the spoil to the side of the trench.

£

Cost per m³ = 3¼ hours at £4.00 — 13.00

Unit rate = £13.00 per m³ (labour £13.00)

Excavate foundation trench, starting at ground level, maximum depth not exceeding 4 m. Per cubic metre (using hand labour)

A labourer will excavate ordinary ground 1 m³ to a depth not exceeding 4 m in about 5 hours, because of the additional lift of loose soil and clearing back the previous excavation from the sides of the trench.*

£

Cost per m³ = 5 hours at £4.00 — 20.00

Unit rate = £20.00 per m³ (labour £20.00)

Excavate basement, starting at ground level, maximum depth not exceeding 2 m deep. Per cubic metre (using hand labour)

A labourer will excavate 1 m³ in ordinary ground in about 3 hours.

£

Cost per m³ = 3 hours at £4.00 — 12.00

Unit rate = £12.00 per m³ (labour £12.00)

Excavate basement, starting at ground level, maximum depth not exceeding 4 m. Per cubic metre (using hand labour)

A labourer will excavate in ordinary ground 1 m³ not exceeding 4 m deep in about 5 hours, because of the additional lift of loose spoil and clearing back sides.*

*Provision of staging or skip loading should also be considered by allowing, say, an extra £1.00 per m³.

	£
Cost per m³ = 5 hours at £4.00	20.00

Unit rate = £20.00 per m³ (labour £20.00)

Excavate foundation trench, starting at ground level, maximum depth not exceeding 2 m. Per cubic metre (using partly machine and partly hand labour)

This is an example of a composite rate for excavation where part of the spoil is won by machine and part by hand labour. It is assumed that the machine will win ⅔ of a cubic metre and hand labour ⅓.

	£
⅔ m³ of excavation to trenches, using a ¼ m³ excavator with banksman at £2.458 per m³ (from previous analysis)	1.639
⅓ m³ of excavation to trenches, using hand labour at £13.00 per m³ (from previous analysis)	4.333
	£5.972

Unit rate = £5.97 per m³ (labour £4.60, plant £1.37)

This item represents the typical problem for an estimator when actually pricing excavation items in bills of quantities. He must use his judgement and experience in assessing the exact rate that will cover him for such factors as the standard bucket being wider or narrower than some or all of the actual trenches, the type of site, the nature of ground, the weather, any delays between excavating and backfilling which may involve partial re-excavation, whether the trenches will be partially obstructed by timber supports, etc. He must be fairly pessimistic in his outlook so that he will be able to cover all these possibilities. The actual excavation may be carried out in optimum conditions where he will make extra profit on his forecast but he must also risk overloading the tender if he is too cautious.

For deeper excavation stages more expensive machines will be necessary and more obstructions likely, particularly timber supports, which will necessitate a higher hand excavation factor.

Remove surplus excavated material from site. Per cubic metre (using hand labour loading into 5 cubic metre lorries, hauling 3 kilometres distance)

When spoil produced by hand excavation is to be disposed in this manner, a temporary spoil heap is formed at the side of the excavation, and the spoil is loaded by hand into the lorry.

A 5 m³ lorry holds 5 m³ of bulked spoil, which represents, say, 4 m³ in the ground. The loading of 5 m³ of loose spoil will take five men 1¼ hours.

		£
Cost of loading 5 m³ = 6¼ hours at £4.00		25.000
Lorry standing loading	75 minutes	
Travelling to tip	15 minutes	
Tipping	10 minutes	
Returning from tip	10 minutes	
Time for 5 m³	110 minutes	
1 hour 50 minutes of a 5 m³ lorry, as previously analysed at £8.360 per hour		15.327
Cost of disposing 5 m³ of loose spoil, or 4 m³ in the ground (exclusive of any tipping charges)		40.327

Cost per m³ $= \frac{£40.327}{4} = \underline{£10.082}$

Unit rate = £10.08 per m³ (labour £6.25, plant £3.83)

Backfill excavated material around foundations. Per cubic metre (using hand labour)

A labourer will backfill and ram 1 m³ in about ¾ of an hour.

	£
Cost per m³ = ¾ hour at £4.00	3.000

Unit rate = £3.00 per m³ (labour £3.00)

Level and trim bottoms of excavation. Per cubic metre (using hand labour)

A labourer will level and trim 6 m² in about 1 hour.

	£
Cost per m² = ⅙ hour at £4.00	0.667

Unit rate = £0.67 per m² (labour £0.67)

All the hand-excavation items analysed so far have been based upon excavation in ordinary ground. If different strata are encountered, the times for ordinary-ground outputs should be multiplied by the following factors:

Ordinary ground:	multiply by	1.00
Heavy clay:	multiply by	1.25
Compact gravel:	multiply by	1.50
Soft chalk:	multiply by	2.50
Hard rock:	multiply by	5.00

Extra over excavating to reduce levels for breaking up 50 mm thick tarmacadam paving with 100 mm thick bed of hardcore below. Per square metre (using hand labour)

The labour output depends upon the condition of the tarmacadam, but under average conditions a labourer will take about 1 hour to break up 1 m² and lay aside for disposal.

	£
1 hour labourer at £4.00 per hour	4.00
This item is to be measured as extra over ordinary excavation therefore the cost of excavating 1 m² 150 mm thick must be deducted. Reduced-level excavation by hand labour as previously analysed = £10.00 per m³.	
Therefore 1 m² 150 mm thick = £10.00 × $\frac{150}{1000}$ =	1.50
Extra cost per m²	2.50

Unit rate = £2.50 per m² (labour £2.50)

Extra over excavating to reduce levels for breaking up 50 mm thick tarmacadam paving with 100 mm thick bed of hardcore below. Per square metre (using a compressor)

A labourer will excavate with compressor equipment about 10 m² of paving in an hour. A compressor with three drills will give an output of 30 m² per hour.

	£
One hour compressor and operator, as previously analysed	5.600
Three drill operators at £4.00 per hour plus allowances in accordance with the NWRA = 3 hours at, say, £4.25	12.750
Cost per hour	18.350

Cost per m² = $\frac{£18.350}{30}$ = £0.612

With mechanical tools, it is the usual practice to discount the excavation in ordinary ground.

Unit rate = £0.61 per m² (labour £0.43, plant £0.18)

Extra over excavating to reduce levels for breaking up 150 mm thick bed of concrete. Per square metre (using hand labour)

A labourer will take about 2½ hours to break up 1 m² and lay aside for disposal.

	£
2½ hours labourer at £4.00 per hour	10.00
Deduct 1 m² 150 mm thick, of reduced-level excavation by hand labour as before = $£10.00 \times \frac{150}{1000}$ =	1.50
Extra cost per m²	£8.50

Unit rate = £8.50 per m² (labour £8.50)

Extra over excavating to reduce levels for breaking up 150 mm thick bed of concrete. Per square metre (using a compressor)

A labourer will excavate, using compressor equipment, about 3¼ m² of 150 mm thick concrete in an hour. A compressor with three concrete-breakers will give an output of, say, 10 m² per hour

	£
One hour of compressor and operator, as previously analysed with three drill operators	18.350

Cost per m² $= \frac{£18.350}{10} = \underline{£1.835}$

With mechanical tools, it is the usual practice to discount the excavation in ordinary ground.

Unit rate = £1.84 per m² (labour £1.28, plant £0.56)

Earthwork support

Earthwork support not exceeding 2 m maximum depth and not exceeding 2 m between opposing faces. Per square metre

Visualize a trench 15 m long, 2 m deep and 1 m wide, i.e. 60 m² of excavation face, with 225 × 38 mm uprights at 1.5 m centres, two 225 × 38 mm runners to each side and two 100 × 100 mm struts between the uprights.
225 × 38 mm planks = 4 × 15 m + 20 × 2 m = 100 m.
100 × 100 mm struts = 20 × 1 m = 20 m.

Cost of timber (assume £140 per m^3)

	m^3	£
100 m of 225 × 38 mm =	0.855	
20 m of 100 × 100 mm =	0.200	
	1.055 m^3 at £140	147.700

Allowing say ten uses, inclusive of waste cost per use =	14.770
A trench 15 m long and 2 m deep will be timbered by two men working 1 day each.	
Sixteen hours labourer at £4.00 per hour plus an allowance for timbering in accordance with the NWRA = 16 hours at say, £4.10	65.600
Two men will strip this timbering in about 3 hours, 6 hours at £4.10	24.600
Cost of timbering 60 m^2	104.970

$$\text{Cost per m}^2 = \frac{£104.970}{60} = \underline{£1.750}$$

Unit rate = £1.75 per m^2 (labour £1.50, material £0.25)

Earthwork support presents similar problems to an estimator as the pricing of excavation. He must use his judgement and experience in assessing an adequate rate bearing in mind the likely depths, the necessity for stronger supports or whether in fact timbering will be required at all. The need for wider excavation to accommodate the members will also have to be considered.

Earthwork support to wider distances between faces of excavation can be built up on similar principles but it is probable that in many cases raking strutting from the base would be required in lieu of the costs of this element being shared between the two sides of a normal trench.

Earthwork support 'left-in' will involve a saving in the 'striking' cost of labour but new timber with one use only must be included in the price build-up. If the builder can be sure that old timber will be available and sufficient he could reduce this cost.

Filling

Broken stone hardcore to pass a 50 mm ring, to make up levels over

250 mm thick, deposited and compacted in layers of 150 mm maximum thickness. Per cubic metre

Cost per tonne of 50 mm gauge broken stone hardcore, delivered to site, say £5.00

	£
Stone hardcore weighs approximately 1.5 tonnes per m^3 therefore 1.5 tonnes at £5.00 =	7.500
Hardcore consolidates approximately 25% when compacted in position, therefore add 25%	1.875
The cost of placing hardcore can vary considerably; under certain conditions, the delivery vehicle can tip the hardcore straight into the final position, or it may be necessary to use a JCB3C type excavator to move and spread the material. However, on confined sites it is possible that the hardcore will have to be tipped and then barrowed by hand into position.	
	9.375
Under average conditions, a labourer will fill about 1 m^3 of consolidated hardcore in an hour.	
1 hour labourer at £4.00 per hour	4.000
	13.375

Unit rate = £13.38 per m^3 (labour £4.00, materials £9.38)

100 mm thick bed of broken stone hardcore to pass a 50 mm ring levelled and compacted. Per square metre

	£
Cost of one m^2 of hardcore 100 mm thick consolidated, from previous analysis = $£9.375 \times \frac{100}{1000} =$	0.938
When placing hardcore in beds, the labourers' time will increase from 1 hour per m^3 to about 1⅓ hours per m^3.	
Cost of labour per m^2, 100 mm thick = $1\frac{1}{3} \times £4.00 \times \frac{100}{1000} =$	0.533
A labourer with a mechanical punner will ram and consolidate about 8 m^2 in an hour. Mechanical punners cost about £1.00 per hour on hire.	
Therefore cost per m^2 = $\frac{£1.00}{8} =$	0.125
Labourer operating punner at £4.00 per hour plus extra allowance in accordance with NWRA,	

say, £4.20 = $\frac{£4.20}{8}$ =	0.525
	2.121

Unit rate = £2.12 per m² (labour £1.05, materials £0.94, plant £0.13)

150 mm thick bed of broken stone hardcore to pass a 50 mm ring levelled and compacted. Per square metre

	£
Cost of 1 m² of hardcore including consolidation	
from previous analysis = £9.375 × $\frac{150}{1000}$ =	1.406
Cost of labour per m² 150 mm thick	
= 1⅓ × £4.00 × $\frac{150}{1000}$ =	0.800
Add consolidation costs from previous analysis	0.125
Labour operating punner as before	0.525
	2.856

Unit rate = £2.86 per m² (labour £1.32, materials £1.41, plant £0.13)

Blinding surface of hardcore with ashes. Per square metre

Cost per tonne of ashes, delivered site, say, £6.00 (there will be no practical finished thickness as the ashes will be consolidated within the general hardcore thickness).

	£
Allow 50 mm thickness of ashes at an approximate weight of 1.5 tonnes per m³ at £6.00 per tonne	0.45
Labourer spreading and blinding by hand 8 m² in about 1 hour.	
Cost per m² = ⅛ hour at £4.00	0.50
	£0.95

Unit rate = £0.95 per m² (labour £0.50, material £0.45)

Note: It will obviously be more economic to use mechanical means of site transport, filling and consolidation whenever possible for all items of excavation and hardcore.

Working space

Excavate working space and filling for foundation trench starting at ground level, maximum depth not exceeding 2 m. Per cubic metre

Note: This is a particularly difficult item to price as SMM6 clause D.12.2 requires an estimator not only to consider the cost of excavation, and whether the allowance is sufficient or not, but also to include for additional earthwork support, disposal of surplus material, surface treatment and all backfilling whether in selected excavation or in other specified materials such as hardcore. The estimator can start by using his previously calculated rate for the main excavation item and add to this his own allowances for the other deemed items but it does not seem to be a satisfactory basis from an estimating viewpoint. Perhaps SMM7 may change working space to a superficial measurement of the areas of the faces of excavation and thereby allow an estimator to make his own allowances.

5 Concrete work

There are three main initial factors to be considered in the cost of concrete:

1 method of mixing
2 materials
3 site transportation and placing

It is necessary for the estimator to make a decision on how the concrete will be mixed as this obviously affects the cost. The main choices available are:

(a) hand
(b) portable machines
(c) 'ready-mix'
(d) batch plant installed on site

The estimator will be influenced by such factors as type of firm, location of site, restrictions on working and storage space, availability of materials including ready-mixed concrete, and the actual quantity and phasing of the concrete required. Mixing by hand is entirely uneconomical and technically unsatisfactory. It takes approximately 5 man hours to measure, water and mix 1 cubic metre of concrete by hand, so the cost is 5 × £4.00 = £20.00.

Concrete is now normally specified by strength rather than giving the proportions of the constituent parts. An estimator must obtain the necessary technical information from a structural adviser which will indicate the materials and proportions necessary to produce concrete to the strength stated.

Cement can be purchased and delivered in a loose state but for normal site use will be transported in paper bags. The delivery price will have to be increased due to the cost of manual unloading and placing into store, for example:

	£
Ordinary Portland cement delivered to site in paper bags, per tonne	£55.00
Off-loading, etc. 1 hour labourer at £4.00 (Continuous work of this nature would justify an extra allowance in accordance with the NWRA)	4.00
Total price per tonne	£59.00

Aggregates are sold by weight or volume depending on particular areas of the British Isles. As the examples shown are based on proportions by volume it will be necessary to convert the costs of the constituents from the weight prices. The average weight of cement is approximately 1420 kg/m^3, sand 1600 kg/m^3, and coarse aggregate 1400 kg/m^3 but the weights for all aggregates will vary according to the particular material and moisture factor.

When the dry materials are mixed together with water, shrinkage of the cement occurs and the fine aggregate and cement fill the spaces between the coarse aggregate particles. This process is assisted by the consolidation of the concrete into its final position. Wastage occurs with the dry materials during site storage and also in transference of the mixed concrete. All these factors can be conveniently grouped together at an approximate arithmetical adjustment of 50%. It is possible for the 'wastage' element to further increase in many ways, for example when trenches have been dug out wider or deeper than the specification because of various site conditions.

The site transportation of mixed concrete and depositing in position in a building can vary to a great extent. A ready-mix lorry could deliver concrete into foundations directly with a minimum of attendant labourers assisting. Sometimes it will be necessary to load mixed concrete into barrows or skips and hoist several floors high.

A more involved procedure of pumping or other specialist means might be necessary for certain contracts. The cost of manual assistance should be included in the price of each item but in most instances any mechanical plant would be included in the preliminaries, or possibly against the plant items in the concrete section.

*Concrete (14 MN/m^2) in trench foundations exceeding 300 mm thick. Per cubic metre (using a 10/7 mixer, continuously)**

		£
Assume a 1:3:6 mix		
Ordinary Portland cement including unloading 1 m^3 at 1420 kg/m^3 at £59.00 per tonne		83.78
Fine aggregate 3 m^3 at 1600 kg/m^3 at £6.00 per tonne		28.80
Coarse aggregate 6 m^3 at 1400 kg/m^3 at £7.00 per tonne		67.20
	c/fwd	£179.78

*A 10/7 mixer takes in 10 cubic feet of dry materials and produces 7 cubic feet of concrete per mix (i.e. 0.20 m^3). Working continuously at the rate of 12 mixes per hour this produces approximately 2.40 m^3 of mixed concrete in an hourly period. This rate, however, would also mean that an output of 96 m^3 would have to be achieved in a 40 hour week to justify the assumption.

	b/fwd	£179.78
Shrinkage, consolidation and wastage *add* 50%		89.89
		269.67

Divide by parts of mix (i.e. 10) $\frac{£269.67}{10}$ = 26.97

Cost of materials in 1 m³ of concrete, as calculation 26.97

The labour gang required for such a machine will be as follows:

1 driver/filler 1 filler 3 wheelers 2 spreaders	7 hours labourer at £4.00 per hour plus an extra 20p per hour for the driver in accordance with the NWRA = £28.20

Cost to produce 2.4 m³ = £28.20

therefore cost for one m³ = $\frac{£28.20}{2.4}$ (but see below) 11.75*

Hourly costs of a 10/7 mixer as previously analysed = £0.765

Mixer produces 2.4 m³ per hour working continuously,

therefore cost per m³ = $\frac{£0.765}{2.4}$ (but see below) 0.32*

39.04

Unit rate = £39.04 per m³ (labour £11.75, materials £26.97, plant £0.32)

Concrete costs vary according to the type of mixer used. Concrete from a 3/2 mixer with an output of 0.06 m³ per mix will be more expensive than that from a 10/7 with an output of 0.20 m³ per mix. This is confirmed by the fact that ready-mixed concrete can be produced at a central mixing plant and transported to sites in costly vehicles for distances up to five kilometres at highly competitive prices.

A concrete mixer working continuously to produce concrete for say a large floor slab will produce more concrete in an hour than the same mixer working intermittently for three separate periods of twenty minutes. This is

* A 10/7 mixer working intermittently would produce approximately 1.6 m³ of concrete in an hour which would result in re-calculations for the above example as follows:

labour gang $\frac{£28.20}{1.6}$ = £17.63

mixing costs $\frac{£0.765}{1.6}$ = £0.48

taken into account when analysing concrete costs, by noting the nature and quantity of the item, and then considering whether the concrete will be produced in continuous or intermittent operations.

In the case of concrete beams and columns which are shown later in this chapter, the volume of concrete would not necessarily justify a mixer operating continuously.

Concrete (14 MN/m²) in trench foundations exceeding 300 mm thick. Per cubic metre (using a 5/3½ mixer, continuously)

A 5/3½ mixer with an output of 0.10 m³ per mix working continuously will produce about 1.2 m³ of concrete in an hour.

	£
Materials for 1 m³ from previous analysis	26.97
The labour gang required for a 5/3½ mixer will be as follows:	

1 driver/filler	5 hours labourer at £4.00 per hour
1 filler	plus an extra 20p per hour for the
2 wheelers	driver in accordance with the
1 spreader	NWRA = £20.20

Labour cost to produce 1.2 m³ = £20.20

$$\text{therefore cost for 1m}^3 = \frac{£20.20}{1.2} \qquad 16.83$$

Hourly costs of a 5/3½ mixer, as previously analysed = £0.474

Mixer produces 1.2 m³ per hour, therefore

$$\text{per m}^3 = \frac{£0.474}{1.2} \qquad 0.40$$

44.20

Unit rate = £44.20 per m³ (labour £16.83, materials £26.97, plant £0.40)

Concrete (14 MN/m²) in trench foundations exceeding 300 mm thick. Per cubic metre (using ready-mixed concrete delivered into position by vehicle)

The costs of placing ready-mixed concrete can vary considerably according to site conditions and the final position of the concrete in the building. For example, under certain conditions, it is possible to have concrete for foundations and floor slabs delivered and tipped from the vehicle into position, so that spreading is the only labour required. On the other hand, there are some sites where the concrete has to be deposited outside the building area and requires barrowing into position.

	£
Cost of 1 m³ of ready-mixed concrete, delivered	35.00
No waste is incurred as the concrete is delivered into position	
A labourer will spread 1 m³ in foundations in about ⅔ of an hour. ⅔ hour labourer at £4.00	2.67
	37.67

Unit rate = £37.67 per m³ (labour £2.67, materials £35.00)

Concrete (14 MN/m²) in trench foundations exceeding 300 mm thick. Per cubic metre (using ready-mixed concrete tipped adjacent to the building and barrowed into position)

	£
Cost of 1 m³ of ready-mixed concrete, delivered	35.00
Residue waste, 2½%	0.88
A labourer will wheel 1 m³ in 1¼ hours including loading into barrows for short distances.	
1¼ hours labourer at £4.00 per hour	5.00
A labourer will spread 1 m³ in foundations in about ⅔ of an hour. ⅔ hour labourer at £4.00	2.67
	43.55

Unit rate = £43.55 per m³ (labour £7.67, materials £35.88)

Concrete (14 MN/m²) in bed not exceeding 100 mm thick. Per cubic metre (using a 5/3½ mixer continuously)

Concrete placed in beds 150 mm or less in thickness will require an additional 2 hours labour for placing over and above the basic price of concrete in foundations.

	£
Basic price of concrete in foundations, using 5/3½ mixer continuously, from previous analysis	44.20
Additional 2 hours labourer at £4.00	8.00
	52.20

Unit rate = £52.20 per m³ (labour £24.83, materials £26.97, plant £0.40)

Concrete (21 MN/m²) in trench foundations exceeding 300 mm thick. Per cubic metre (using a 5/3½ mixer continuously)

	£
Assume a 1:2:4 mix.	
Ordinary Portland cement, 1 m³ at 1420 kg/m³ at £59.00 per tonne	83.78
Fine aggregate 2 m³ at 1600 kg/m³ at £6.00 per tonne	19.20
Coarse aggregate 4 m³ at 1400 kg/m³ at £7.00 per tonne	39.20
	142.18
Consolidation and waste *add* 50%	71.09
	213.27
Divide by parts of mix (i.e. 7) $\frac{£213.27}{7}$ =	30.47
A 5/3½ mixer working continuously will produce about 1.2 m³ of concrete in an hour.	
Cost of labour gang working a 5/3½ mixer continuously, from previous analysis = £20.20, therefore cost per m³ = $\frac{£20.20}{1.2}$ =	16.83
Hourly cost of 5/3½ mixer, from previous analysis, = £0.474, therefore cost per m³ = $\frac{£0.474}{1.2}$ =	0.40
	47.70

Unit rate = £47.70 per m³ (labour £16.83, materials £30.47, plant £0.40)

Concrete (21 MN/m²) in bed not exceeding 100 mm thick. Per cubic metre (using a 5/3½ mixer continuously)

Concrete placed in beds 150 mm or less in thickness will require an additional 2 hours labour for placing, over and above the basic price of concrete in foundations.

	£
Basic price of 1:2:4 mix concrete in foundations, from previous analysis, 5/3½ mixer, continuous use	47.70
Additional labour 2 hours at £4.00	8.00
	55.70

Unit rate = £55.70 per m³ (labour £24.83, materials £30.47, plant £0.40)

Concrete ($21\ MN/m^2$) in bed not exceeding 100 mm thick worked around mesh reinforcement. Per cubic metre (using a 5/3½ mixer continuously)

	£
Basic price of concrete 1:2:4 mix in foundations as before	47.70
Additional labour for beds, 2 hours at £4.00	8.00
Additional labour in working concrete around reinforcement, 1 hour per m^3 at £4.00	4.00
	59.70

Unit rate = £59.70 per m^3 (labour £28.83, materials £30.47, plant £0.40)

Vibrating

If a poker vibrator is used to ensure better consolidation around the reinforcement, the cost would have to be added to the above example as follows:
Assume capital and running costs of the vibrator to be £1.50 per hour.

$$\text{Cost per } m^3 = \frac{£1.50}{1.2} = \underline{£1.25} \text{ per } m^3$$

No additional labour would be required as the previously allotted spreading time would be partially allocated to the vibrator operator.

Reinforced concrete ($21\ MN/m^2$) in walls exceeding 100 mm and not exceeding 150 mm thick. Per cubic metre (using a 5/3½ mixer continuously)

Concrete placed in formwork in walls 150 mm thick will require about 6½ hours additional labour per m^3 over and above the basic price of concrete in foundations.

	£
Basic price of concrete per m^3 in foundations, using a 5/3½ mixer continuously (as before)	47.70
Additional labour, placing in 150 mm thick walls, 6½ hours at £4.00 per hour	26.00
Additional labour in working concrete around reinforcement 1 hour at £4.00	4.00
	77.70

Unit rate = £77.70 per m^3 (labour £46.83, materials £30.47, plant £0.40)

Reinforced concrete (21 MN/m²) in walls exceeding 300 mm thick. Per cubic metre (using a 5/3½ mixer continuously)

Concrete placed in formwork in walls over 300 mm thick will require about 4 hours additional labour per m³ over and above the basic price of concrete in foundations

	£
Basic price of concrete in foundations, 5/3½ mixer used continuously (as before)	47.70
Additional labour, placing in walls, 4 hours at £4.00 per hour	16.00
Additional labour in working concrete around reinforcement, 1 hour at £4.00	4.00
	67.70

Unit rate = £67.70 per m² (labour £36.83, materials £30.47, plant £0.40)

Reinforced concrete (21 MN/m²) in roof slab not exceeding 100 mm thick. Per cubic metre (using a 5/3½ mixer continuously)

Concrete hoisted and filled into formwork in 100 mm thick roof slabs requires about 5¼ hours additional labour per m³ over and above the basic price of concrete foundations.

	£
Basic price of concrete in foundations, 5/3½ mixer used continuously (as before)	47.70
Additional labour, placing in 100 mm thick roof slab, 5¼ hours at £4.00 per hour	21.00
Additional labour, working around reinforcement, 1 hour at £4.00	4.00
	72.70

Unit rate = £72.70 per m³ (labour £41.83, materials £30.47, plant £0.40)

Reinforced concrete (21 MN/m²) in isolated beams not exceeding 0.03 square metres sectional area. Per cubic metre (using a 5/3½ mixer continuously)

Concrete hoisted and filled into formwork in beams not exceeding 0.03 m² sectional area requires about 10 hours additional labour per m³ over and above the basic price of concrete in foundations.

	£
Basic price of concrete in foundations, with a 5/3½ mixer used continuously (as before)	47.70
Additional labour, placing in beams not exceeding 0.05 m³ sectional area, 10 hours at £4.00 per hour	40.00
Additional labour in working around reinforcement, 1 hour at £4.00 per hour	4.00
	91.70

Unit rate = £91.70 per m³ (labour £60.83, materials £30.47, plant £0.40)

Reinforced concrete (21 MN/m²) in isolated beams exceeding 0.03 but not exceeding 0.10 square metres sectional area. Per cubic metre (using a 5/3½ mixer continuously)

Concrete hoisted and filled into formwork in beams exceeding 0.03 but not exceeding 0.10 m² sectional area requires about 8 hours additional labour per m³ over and above the basic price of concrete in foundations.

	£
Basic price of concrete in foundations, 5/3½ mixer used continuously (as before)	47.70
Additional labour placing in beams, 8 hours at £4.00 per hour	32.00
Additional labour working around reinforcement, 1 hour at £4.00 per hour	4.00
	83.70

Unit rate = £83.70 per m³ (labour £52.83, materials £30.47, plant £0.40)

Reinforced concrete (21 MN/m²) in isolated columns not exceeding 0.03 square metres sectional area. Per cubic metre (using a 5/3½ mixer continuously)

Concrete filled into formwork in columns not exceeding 0.03 m² sectional area requires about 12 hours additional labour per m³ over and above the basic price of concrete in foundations.

	£
Basic price of concrete in foundations (as before)	47.70
Additional labour placing in columns, 12 hours at £4.00 per hour	48.00
Additional labour working around reinforcement, 1 hour at £4.00 per hour	4.00
	99.70

Unit rate = £99.70 per m^3 (labour £68.83, materials £30.47, plant £0.40)

Steel wire square-mesh fabric reinforcement to comply with BS 4483 Reference A 142, weighing 2.22 kg per square metre, in suspended slabs with 150 mm minimum laps at all joints, per square metre

	£
Cost to purchase 1 m^2 of fabric in 5 × 2 m sheets at 90p per m^2	0.900
Allowance for laps, 10%	0.090
Allowance for waste, 5%	0.045
Allow for tying-wire and distance blocks, say	0.100
Offloading, two labourers 5 minutes each per sheet = 10 minutes labourer's time = £0.667 for 10 m^2, therefore cost per m^2 = say, 7p	0.070
Two labourers will fix about 20 m^2 of this weight of fabric in roofs in 1 hour. Therefore 2 hours at £4.00 (see WRA) per hour = £8.00. Cost to fix 20 m^2 £8.00, therefore cost for 1 m^2 = $\frac{£8.00}{20}$ = £0.40	0.400
	1.605

Unit rate = £1.61 per m^2 (labour £0.47, materials £1.14)

Steel wire square-mesh fabric reinforcement to comply with BS 4483 Reference A 193, weighing 3.02 kg per square metre, in suspended slabs with 150 mm minimum laps at all joints, per square metre

	£
Cost of 1 m^2 of fabric	1.20
Allowance for laps, 10%	0.12
Allowance for waste, 5%	0.06
Allow for tying-wire and distance blocks, say	0.10
Offloading fabric (as before) at 7p per m^2	0.07
Two labourers will fix about 16 m^2 of this weight of fabric in roofs in 1 hour = £8.00	
Cost per m^2 = $\frac{£8.00}{16}$ =	0.50
	2.05

Unit rate = £2.05 per m^2 (labour £0.57, materials £1.48)

12 mm diameter mild steel bar reinforcement in isolated beams. Per tonne

	£
Cost of 1 tonne of 12 mm diameter bars, delivered site	400.00
Allowance for waste in cutting bars to length, 10%	40.00
c/fwd	£440.00

	£
b/fwd	440.00
Annealed tying-wire, say 10 kg per tonne at £3.00 per kg	30.00
Offloading and stacking: 1 tonne will take a labourer about 3 hours to offload.	
3 hours labourer at £4.00 per hour	12.00
It will take a skilled steel-reinforcement fixer about 60 hours to cut, bend, and fix 1 tonne of 12 mm steel bars in beams. Qualified steel-reinforcement fixers are paid the craftsman's rate, in accordance with the NWRA.	
60 hours at £5.00 per hour	300.00
Cost per tonne =	782.00

Unit rate = £782.00 per tonne (labour £312.00, materials £470.00)

12 mm diameter mild steel bar reinforcement in beams, all as last described, but using pre-formed bars. Per tonne

	£
Cost of 1 tonne of 12 mm bars, delivered to site cut, bent bundled and labelled	450.00
Nominal allowance for waste 2½%	11.25
Tying-wire as before	30.00
Offloading as before	12.00
It will take a skilled fixer about 40 hours to assemble and fix 1 tonne of 12 mm pre-formed steel bars in beams.	
40 hours at £5.00	200.00
Cost per tonne =	703.25

Unit rate = £703.25 per tonne (labour £212.00, materials £491.25)

10 mm diameter mild steel bar reinforcement in isolated beams. Per tonne

	£
Cost of 1 tonne of 10 mm diameter bars, delivered site	420.00
Allowance for waste in cutting to length, 10%	42.00
Annealed tying-wire, say 14 kg per tonne at £3.00 per kg, including waste	42.00
Offloading and stacking (as before)	12.00
It will take a skilled reinforcement fixer about 90 hours to cut, bend and fix 1 tonne of 10 mm diameter bars.	
90 hours steel fixer at £5.00 per hour	450.00
Cost per tonne =	966.00

Unit rate = £966.00 per tonne (labour £462.00, materials £504.00)

6 mm diameter mild steel bar reinforcement in stirrups to isolated beams. Per tonne

	£
Cost of 1 tonne of 6 mm diameter bars, delivered site	450.00
Allowance for waste in cutting to length, 10%	45.00
Annealed tying-wire, say 20 kg per tonne at £3.00 per kg	60.00
Offloading and stacking (as before)	12.00
It will take a skilled reinforcement fixer about 130 hours to cut, bend and fix 1 tonne of 6 mm diameter bars in stirrups, as there are about four times more 6 mm diameter bars to the tonne than 12 mm, and stirrups are in short lengths.	
130 hours at £5.00	650.000
Cost per tonne =	1217.00

Unit rate = £1217 per tonne (labour £662.00, materials £555.00)

Note: Appropriate adjustments, as previously illustrated, should be made to the last two examples if the bars are purchased pre-formed.

Formwork

Formwork is carried out by carpenters or joiners (the term varies in different parts of the country).

One of the major principles of estimating costs of formwork is to accurately determine the number of uses that will occur. If a single flight *in situ* concrete staircase is being constructed, it is very likely that new timber will be used and virtually thrown away after one use. In this case the cost of formwork will be far higher than the combined costs of the concrete and reinforcement because of not only the formwork materials but also the tradesman's time involved in making and erecting the shuttering. This example clearly illustrates that the use of standard precast concrete components is very desirable for stairs, lintels, floor slabs, etc. wherever possible not only on economic grounds but also for speed of construction. On the other hand if a building contained many *in situ* concrete columns it would be very economical for the contractor to purchase lengths of special metal sections that could be used many times on several contracts. Alternatively plywood-faced sections could be constructed for columns, beams, walls, floors, etc. bolted together in position and re-used on at least several occasions. The finish required on the concrete would normally be specified in the preambles and referred to in the item description, for example, rough, fair, smooth, etc. and the design of the lining to the formwork would then have to be carefully considered (SMM6 clause F.13.9).

The following examples show traditional ideas of building up costs but it has become obvious on many sites that wastage of timber in formwork has increased in recent years by not re-using the materials as often as possible.

Formwork to horizontal soffit of slab, in four separate surfaces. Per square metre

The amount of strutting timber required varies according to the height of the soffit above floor level. The Standard Method of Measurement recognizes this by requiring the separation of formwork to soffits over 3.5 m high, grouped in further stages of 1.5 m. Consider an area of 10 m² using 40 mm thick boarding with shuttering timber at £150 per m³ (unloaded). The estimator would refer specifically to the drawings.

	£
The cost of 1 m² of 40 mm boarding is	
$£150 \times \frac{40}{1000} = £6.00$	
10 m² at £6.00 per m² =	60.00
For soffits not exceeding 3.5 m high assume an amount of joists and strutting timber required is about 0.60 m³ per 10 m², but in practice this would have to be designed and calculated.	
0.60 m³ at £150 per m³ =	90.00
	150.00
Timber for formwork with care and good management, can be used about five times before becoming unusable.	
Therefore cost per use $= \frac{£150.00}{5} =$	30.00
Waste on each use, say 10%	3.00
Nails, bolts, and mould oil, say 30p per m², 10 m² at 30p =	3.00
It will take two joiners, assisted by one labourer, about 8 hours to erect and dismantle this formwork to a soffit of 10 m²	
16 hours joiner at £5.00 per hour	80.00
8 hours labourer at £4.00 per hour	32.00
	148.00

Total cost for 10 m² = £148.00

Cost per m² $= \frac{£148.00}{10} = \underline{£14.80}$

Unit rate = £14.80 per m² (labour £11.20, materials £3.60)

Formwork to attached horizontal beams size 200 × 300 mm in five members. Per linear metre

Consider a typical beam by reference to the drawings and possibly a bill diagram, say 10 m long. Therefore the girth requiring support = 2 × 300 mm + 200 mm = 800 mm and 8 m² for the total beam.

	£
8 m² of 18 mm plywood at £6.00 per m²	48.000
A high allowance for cutting will be required, say 15%	7.200
For sides and soffits of beams assume that 0.50 m³ of props and struts, etc. are required for 8 m² of surface to be supported. 0.50 m³ timber at £150 m³	75.000
	130.200
Allowing five uses (as before) cost per use	26.040
Waste on each use, say 10%	2.604
Nails, bolts and mould oil, say 30p per m² × 8 m²	2.400
It will take two joiners assisted by one labourer, about 8 hours to erect and dismantle this formwork to beams	
Joiner, 16 hours at £5.00 per hour	80.000
Labourer, 8 hours at £4.00 per hour	32.000
	143.044

Total cost per 10 m = £143.044

$$\text{Cost per m} = \frac{£143.044}{10} = \underline{£14.304}$$

Unit rate = £14.30 per m (labourer £11.20, materials £3.10)

Formwork (smooth finish to be left on concrete as specified) to attached horizontal beams size 200 × 300 mm in five members. Per linear metre

	£
Cost per use of materials, as previous item	26.040
Waste on each use, say 10%	2.604
Nails, bolts and mould oil, say 30p per m² × 8m² (as before)	2.400
Because of the finish required the labour gang will take a little longer, as small surface defects have to be made good. It will take two joiners, accompanied by one labourer, say 10 hours.	
Joiner, 20 hours at £5.00 per hour	100.000
Labourer, 10 hours at £4.00 per hour	40.000
	171.044

$$\text{Cost per m} = \frac{£171.044}{10} = \underline{£17.104}$$

Unit rate = £17.10 per m (labour £14.00, materials £3.10)

Note: For some finishes on concrete it may be necessary for a plasterer to rub up and smooth the finished surface after the formwork has been 'struck'.

Formwork to isolated columns size 300 × 300 mm in ten members. Per linear metre

Consider a typical column by reference to the drawings and possibly a bill diagram, say, 3 m long. Therefore the girth requiring support = 4 × 300 mm = 1200 mm and 3.60 m² area for the column.

	£
3.60 m² of 18 mm plywood at £6.00 per m²	21.60
Cutting waste, etc. 15%	3.24
Allow 0.20 m³ of timber supports at £150 per m³ =	30.00
	54.84
Allowing five uses for timber (as before)	10.97
Waste on each use, say 10%	1.10
Nails, bolts and mould oil, say 30p per m² × 3.6 m²	1.08
The labour in erecting formwork to columns will be similar to beams.	
Joiner, 8 hours at £5.00 per hour	40.00
Labourer, 4 hours at £4.00 per hour	16.00
	69.15

$$\text{Cost per m} = \frac{£69.15}{3} = \underline{£23.05}$$

Unit rate = £23.05 (labour £18.67, materials £4.38)

Concrete (21 MN/m²) in precast lintel 215 × 150 mm × 1 m long, reinforced with and including No. 2, 12 mm diameter mild steel bars, left rough for plaster, bedded in cement mortar 1 – 3 mix. Each

A 215 × 150 mm lintel 1 m long requires 0.032 m³ of concrete.

		£
For this analysis consider 31 lintels each 1 m long, 1 m^3 of concrete 1:2:4 mix 19 mm aggregate mixed with a 5/3½ mixer used continuously, and placed in foundations as before =		47.70
An additional 4 hours labour will be required, over and above the labour for foundation concrete for placing in small box moulds.		
4 hours labourer at £4.00 per hour		16.00
A 1 m long lintel will require bars say 1.2 m long or 2.4 m of 12 mm diameter bar for each of 31 lintels = 74.4 m at 0.888 kg per m = 66 kg at £782.00 per tonne as before cut and placed in beams		51.61
	£	
A 1 m long lintel requires 0.70 m^2 of 40 mm thick boarding at £6.00 per m^2 from previous analysis =	4.20	
Mould oil and bolts say 60 p =	0.60	
Joiner making box ½ hour at £5.00 =	2.50	
Cost of one box =	7.30	
With say, twenty uses per box, the cost per use = $\frac{£7.30}{20}$ = £0.365		
Box used for 31 lintels = 31 × £0.365 =		11.32
		126.63
Cost of casting 31, 1 m long lintels = £126.63		
Cost per linear m = $\frac{£126.63}{31}$ =		4.08
The cost to cast any shaped lintel can be calculated as a *pro-rata* rate using the details given above as a basis.		
It will take two labourers 10 minutes each to take a lintel from stack when required and to hoist on to scaffold, ready for walling in.		
20 minutes labourer at £4.00 per hour		1.67
It will take a bricklayer and labourer about 15 minutes to lift up the lintel from the scaffold, and bed and level in position.		
¼ hour bricklayer at £5.00 per hour		1.25
¼ hour labourer at £4.00 per hour		1.00
		8.00

Unit rate = £8.00 each (labour £4.83, materials £3.17)

Concrete (21 MN/m²) in precast lintel 215 × 150 mm × 1 m long, reinforced with and including No. 2, 12 mm diameter mild steel bars left rough for plaster, bedded in cement mortar 1:3 mix. Each (using a precast concrete lintel supplied by a precast-concrete manufacturer)

	£
Consider a 1 m concrete lintel.	
Cost to purchase, say	5.00
Allow a nominal waste 2½%	0.13
Take the offloading and fixing costs as the previous example	3.92
	9.05

Unit rate = £9.05 each (labour £3.92, materials £5.13)

It is difficult to compare the economic costs of purchasing precast concrete lintels as opposed to precasting them on site. Some contractors contend that any available surplus concrete should be used to make the odd lintels on site. The authors cannot accept this as being economical or efficient and consider that such members should be cast in a specialist environment.

Precast pre-stressed concrete proprietary floor slab of beams size 300 mm wide × 150 mm deep × 8 m long. Each

	£
Floor slab beams delivered site in 8 m lengths	48.00
Nominal wastage allowance 2½%	1.20
Assume that a mobile crane has to be hired specially for unloading and fixing this particular item with a minimum hire period of ½ day. It will be necessary for 6 labourers to assist with the operation. An area of 100 m² will be hoisted and set in position in about 4 hours (assume 40 beams).	
Hire of crane and operator, etc. at £20.00 per hour $= \frac{£20.00 \times 4}{40}$ = cost per beam	2.00
Labour $\frac{6 \times 4 \times £4.00}{40}$ = cost per beam	2.40
	53.60

Unit rate = £53.60 each (labour £2.40, materials £49.20, plant £2.00)

The costs of grouting joints, etc. would also have to be considered.

Note: The plant costs for this item would be a good example for setting against the plant items in the concrete section of the bill of quantities. The hirer might charge separately for transport to and from the site but in this case it is unlikely.

6 Brickwork and blockwork

Mortars

Mortars can be cement and sand, or cement, lime, and sand. They can be entirely mixed on site or delivered in a plastic state as sand and lime with cement added when required. In recent years the use of masonry cements and plasticizers has tended to become more widespread to the exclusion of lime.

The wastage of dry materials, shrinkage and general mixed mortar wastage can be conveniently grouped together as approximately 33⅓ per cent. This figure is a reduction from the 50 per cent factor indicated for concrete using graded aggregates as there are no coarse aggregate particles which collect the cement and fine aggregate between them.

Cement mortar 1:3 mix per cubic metre

The labour costs involved in the mixing of mortar by machine are accounted for in the hourly cost of the bricklaying gang. This means that the labourers in the gang mix the mortar as required from time to time, as well as keeping the bricklayers supplied with bricks.

	£
Ordinary Portland cement including unloading 1 m^3 at 1420 kg/m^3 at £59.00 per tonne	83.78
Building sand 3 m^3 at 1600 kg/m^3 at £7.00 per tonne	33.60
	117.38
Shrinkage, waste, etc. *add* 33⅓%	39.13
	156.51
Divide by parts of mix (i.e. 4) $= \dfrac{£156.51}{4} =$	39.13
Use of 5/3½ mixer for 1 hour, which is the time taken to produce 1 m^3 of mortar	0.47
Cost per m^3	39.60

Common bricks: Quantities of bricks and mortar

	Bricks	*Mortar*	
215 × 102.5 × 65 mm bricks with 10 mm joints	*Number of bricks*	*No frog* (cu m)	*One frog* (cu m)
Half-brick wall per square metre	59	0.02	0.03
One-brick wall per square metre	118	0.05	0.07
One-and-a-half-brick wall per square metre	177	0.08	0.11

Facing bricks per square metre

Brickwork bond Joints 10 mm wide	*Brick sizes* 215 × 102.5 × 65 mm
English bond	89
Flemish bond	79
English garden wall bond	74
Flemish garden wall bond	68
Stretcher bond	59
Header bond	118

	£
Cement and sand plain mortar (from previous analysis)	39.60
Plasticizer, allow 3 litres at £1.00 per litre	3.00
	42.60

Cost per m³ = £42.60

Cement, lime, sand, gauged mortar 1:1:6 mix. Per cubic metre

	£
Cement 1 m³ (as before)	83.78
Bagged hydrated lime including unloading 1 m³ at 700 kg/m³ at £80.00 per tonne	56.00
Building sand 6 m³ at 1600 kg/m³ at £7.00 per tonne	67.20
	206.98
Shrinkage, waste, etc. *add* 33⅓ %	68.99
	275.97

Divide by parts of mix (i.e. 8) $= \dfrac{£275.97}{8} =$ 34.50

Use of 5/3½ mixer for 1 hour to mix 1 m³ of mortar
(as previously analysed) 0.47

Cost per m³ = 34.97

Common bricks

The cost of common bricks will vary greatly according to the delivery area and is dependent generally on haulage costs. Wastage of bricks is directly related to site efficiency in matters of storage, over-ordering, etc. Labour gang ratios and outputs are affected by bonus schemes, nature and position of work, type of bricks and mortar, and the degree of cutting involved, etc.

One-brick-thick (215 mm) wall in 65 mm common bricks in foundations, in cement mortar 1:3 mix, laid English bond. Per square metre

	£
118 bricks are required to wall one square metre 215 mm thick. 118 bricks at £65.00 per thousand, delivered site, tipped $= \dfrac{£65.00}{1000} \times 118$	7.670
Waste on bricks (*note:* bats can be used), 5%	0.384
65 mm bricks with one frog require 0.07 m³ of mortar per m² 215 mm thick. 0.07 m³ at £39.60 per m³	2.772

The labour gang required for walling brickwork in foundations will comprise 2 bricklayers and 1 labourer (a 2:1 gang). This is because materials are handled once, the bricks being stacked and the mortar boards placed at ground level.

Cost of gang for 1 hour:	£
2 hours bricklayer at £5.00 per hour	10.00
1 hour labourer at £4.00 per hour	4.00
	14.00

Rough-walling at ground level, a bricklayer will lay about 60 bricks per hour. Gang will lay 120 bricks per hour

Cost of laying 118 bricks $= \dfrac{£14.00}{120} \times 118$ 13.767

24.593

Unit rate = £24.59 per m² (labour £13.77, materials £10.82)

One-brick-thick (215 mm) wall in 65 mm fletton bricks in foundations, in cement mortar 1:3 mix, laid English bond. Per square metre

65 mm London Brick Company flettons are delivered to the site in lorry loads of 6000 bricks, which have to be offloaded by hand and roughly stacked. Whereas it will take about 1½ man hours to offload and stack 1000 facing bricks, 1000 flettons can be offloaded and roughly stacked by a labourer in about 1 hour.
Flettons delivered site cost £60.00 per 1000 plus
  1 hour labourer offloading and stacking at £4.00
  per hour = £64.00

	£
Cost for 118 bricks $= \frac{\pounds 64.00}{1000} \times 118 =$	7.552
Waste on bricks 5%	0.378
0.07 m^3 of mortar at £39.60 per m^3	2.772
Labour cost for laying 118 bricks (from previous analysis)	13.767
	24.469

Unit rate = £24.47 per m^2 (labour £13.77, materials £10.70)

Half-brick-thick (103 mm) skin of hollow wall in 65 mm common bricks in foundations, in cement mortar 1:3 mix, laid stretcher bond. Per square metre

	£
Half-brick walls in 65 mm bricks require 59 bricks per m^2. 59 No. 65 mm bricks at £65.00 per thousand, tipped	3.835
Waste on bricks, 5%	0.192
65 mm bricks with one frog, when walled half-brick thickness, require 0.03 m^3 at £39.60 per m^3	1.188
In practice, there is no difference in the labour output when walling half-brick skins of hollow walls and when walling solid walls. However, if strict site supervision requires the clearing of the cavity with a cavity lath, a bricklayer's output will fall from 60 bricks per hour to about 50.	
Cost of laying 59 bricks $= \frac{\pounds 14.00}{120} \times 59$	6.883
	12.098

Unit rate = £12.10 per m^2 (labour £6.88, materials £5.22)

Half-brick-thick (102 mm) honeycombed sleeper wall in 65 mm common bricks, in cement mortar 1:3 mix, laid stretcher bond. Per square metre

About 45 No. 65 mm bricks will be required per m² for half-brick sleeper walls, owing to the voids.

	£
45 common bricks at £65.00 per thousand, tipped $= \frac{£65.00}{1000} \times 45$	2.925
Waste on bricks, 5%	0.146
A normal half-brick wall requires 0.03 m³ of mortar; for half-brick honeycombed mortar requirements, say 0.02 m³ at £39.60 per m³	0.792
Cost of laying 45 bricks $= \frac{£14.00}{120} \times 45$	5.250
	9.113

Unit rate = £9.11 per m² (labour £5.25, materials £3.86)

Form a 50 mm wide cavity to a hollow wall, with and including four galvanized twisted-pattern wall-ties. Per square metre

	£
Galvanized twisted-pattern wall-ties, cost 7p each delivered site. Cost of four ties =	0.280
Waste on ties for loss on site, 10%	0.030
As mentioned previously, unless the cavity is to be kept clear by means of a cavity lath the brickwork output does not fall, therefore there is no additional labour for forming a cavity. If the cavity is to be kept clear, the bricklayer's output will fall from 60 to 50 bricks per hour; this will represent the labour cost of forming the cavity. Allow, say 5 minutes per m² (bricklayer only) $= \frac{£5.00}{12} =$	0.420
	0.730

Unit rate = £0.73 per m² (labour £0.42, materials £0.31)

One-brick-thick (215 mm) wall in 65 mm common bricks in superstructure, in cement mortar 1:3 mix, laid English bond. Per square metre

	£
118 No. 65 mm common bricks at £65.00 per thousand (as before)	7.670
Waste on bricks, 5% (as before)	0.384
0.07 m³ of cement mortar at £39.60 per m³ (as before)	2.772
c/fwd	£10.826

	b/fwd £10.826
The labour gang required for walling brickwork in superstructures is different from that required for foundation work. Because of the additional hoisting and working from a scaffold, the gang will consist of 3 bricklayers and 2 labourers (a 3:2 gang).	£
3 hours bricklayer at £5.00 per hour	15.00
2 hours labourer at £4.00 per hour	8.00
	23.00
Walling off a scaffold: since superstructure work is generally of a better quality, a bricklayer's output will fall from 60 to 50 bricks per hour. Therefore the gang will lay 150 bricks per hour.	
Cost of laying 118 bricks $= \dfrac{£23.00}{150} \times 118$	18.093
	28.919

Unit rate = £28.92 per m² (labour £18.09, materials £10.83)

One-and-a-half-brick-thick (328 mm) wall, in 65 mm common bricks in superstructure, in cement mortar 1:3 mix, laid English bond. Per square metre

	£
177 No. 65 mm common bricks at £65.00 per thousand, delivered site, tipped	11.505
Waste on bricks (as before), 5%	0.575
0.11 m³ of cement mortar at £39.60 per m³	4.356
Cost of laying 177 bricks at output of 150 bricks per hour $= \dfrac{£23.00}{150} \times 177$	27.140
	43.576

Unit rate = £43.58 per m² (labour £27.14, materials £16.44)

Half-brick-thick (103 mm) skin of hollow wall, in 65 mm common bricks in superstructure, in cement mortar 1:3 mix, laid stretcher bond. Per square metre

	£
59 No. 65 mm bricks at £65.00 per thousand delivered site, tipped (as before)	3.835
Waste on bricks, 5%	0.192
0.03 m³ of mortar at £39.60 per m³ (as before)	1.188
Cost of laying 59 bricks at 150 bricks per hour $\dfrac{£23.00}{150} \times 59$	9.047
	14.262

Unit rate = £14.26 per m² (labour £9.05, materials £5.21)

Brickwork one-brick-thick (215 mm) in projections, in 65 mm common bricks in superstructure, in cement mortar 1:3 mix, laid English bond. Per square metre

	£
118 No. 65 mm common bricks at £65.00 per thousand, delivered site, tipped (as before)	7.670
Waste on bricks, 5%	0.384
0.07 m^3 of mortar at £39.60 per m^3	2.772
With the extra cutting entailed when walling in projections to walls, a bricklayer's output will fall from 50 to about 45 bricks per hour. Therefore the gang's output will be 135 bricks per hour.	
Cost of laying 118 bricks $= \frac{£23.00}{135} \times 118$	20.104
	30.930

Unit rate = £30.93 per m^2 (labour £20.10, materials £10.83)

Brickwork one-brick-thick (215 mm), in isolated piers in 65 mm common bricks in superstructure, in cement mortar 1:3 mix, laid English bond. Per square metre

	£
118 No. 65 mm common bricks at £65.00 per thousand, delivered site, tipped (as before)	7.670
Waste on bricks (as before), 5%	0.384
0.07 m^3 of mortar at £39.60	2.772
Owing to the isolated nature of this work and the fact that it requires far more plumbing up of angles, a bricklayer's output will fall from 50 bricks per hour to about 35. Therefore the gang will lay 105 bricks per hour.	
Cost of laying 118 bricks $= \frac{£23.00}{105} \times 118$	25.848
	36.674

Unit rate = £36.67 per m^2 (labour £25.84, materials £10.83)

Extra over common brickwork for selected fair face and pointing with a weather joint. Per square metre

No additional material is required for this item as the bricks and mortar have already been allowed for in the main item. A fall in output from the bricklayer is involved because of the greater care necessary in setting the face and pointing the joints.

	£
Allow an extra ½ hour per m^2 bricklayer only at £5.00 per hour	2.50

Unit rate = £2.50 per m^2 (labour £2.50)

Bed wall-plate in cement mortar 1:3 mix. Per linear metre

A bricklayer will bed about 12 m of wall-plate in one hour; as this is carried out concurrently with the walling, a labourer should be included in the cost (3:2 ratio).

	£
Mortar required = 12 m × 103 mm × 10 mm = 0.012 m^3 at £39.60 per m^3 (from previous analysis)	0.475
1 hour bricklayer and attendant labourer at £7.667 per hour	7.667
	8.142

$$\text{Cost per m} = \frac{£8.142}{12} = \underline{0.679}$$

Unit rate = £0.68 per m (labour £0.64, materials £0.04)

Bed door-frame and point externally with an oiled mastic fillet. Per linear metre

The frames are walled in with the general brickwork; this item resolves into the pointing up in mastic later. The cement mortar is disregarded, as this is included with the general walling. A bricklayer will mastic-point all frames after the walling is completed, and therefore does not require the services of an attendant labourer.

Visualize a door-frame about 5 m girth.

	£
Mastic for 5 m at 10p per m	0.50
It will take a bricklayer about quarter of an hour with a gun to point up such a frame. ¼ hour bricklayer at £5.00 per hour	1.25
	1.75

$$\text{Cost per m} = \frac{£1.75}{5} = \underline{£0.35}$$

Unit rate = £0.35 per m (labour £0.25, materials £0.10)

Close 50 mm cavity at jamb of opening with cut brickwork. Per linear metre

This is another item which is difficult to visualize, as the bricklayer carries out the operation as he builds the wall. However, it is reasonable to assume that a bricklayer closes about 3 metres of cavity in one hour.

	£
Approximately 4 extra bricks are required as bats per m, therefore for 3 m, 12 bricks are required at £65.00 per thousand $\frac{£65.00}{1000} \times 12$	0.780
As this operation is carried out simultaneously with the walling, the attendant labourer must be included. 1 hour bricklayer and attendant labourer at £7.667 per hour (3:2 ratio)	7.667
	8.447

Cost per m $= \frac{£8.447}{3} = \underline{£2.816}$

Unit rate = £2.82 per m (labour £2.56, materials £0.26)

Metal-mesh reinforcing fabric 24-gauge (0.56 mm) 64 mm wide, built into bed joint of brickwork. Per linear metre

	£
An 82 m coil of this mesh cost £12.00	12.00
Waste, 5%	0.60
Laps and passings, 5%	0.60
This fabric is laid during the general walling, therefore a labourer's time must be included. It will take a bricklayer about 4 hours to lay 82 m of the fabric, owing to the frequent cutting at angles and the difficulty of walling bricks on top of the fabric.	
Bricklayer and attendant labourer, 4 hours at £7.667 per hour (3:2 ratio)	30.67
	43.87

Cost per m $= \frac{£43.87}{82} = \underline{£0.535}$

Unit rate = £0.54 per m^2 (labour £0.37, materials £0.17)

215 × 65 mm red terracotta air brick. Each

	£
Air bricks cost to purchase 75p each	0.750
Waste on bricks for breakage and loss, 10%	0.075
Allow an extra 3 minutes at £7.667 (3:2 ratio)	0.383
	1.208

Unit rate = £1.21 per m² (labour £0.38, materials £0.83)

215 × 215 mm red terracotta air brick, walled into outer skin of cavity wall, form opening in inner skin, and line out opening in cement mortar. Each

	£
Air bricks cost £1.00 each	1.00
Waste, 10%	0.10
Mortar, say 5p	0.05
Fixing costs: bricklayer with attendant labourer (3:2 ratio) 10 minutes at £7.667	1.28
	2.43

Unit rate = £2.43 each (labour £1.28, materials £1.15)

Rake out joint of brickwork for horizontal lead flashing, and point in cement mortar 1:3 mix. Per linear metre

A bricklayer will rake out about 6 m of joint in one hour; no attendant labourer is required, as this work is not usually done with the general walling.

	£
1 hour bricklayer raking out joint at £5.00	5.00
A bricklayer will point about 10 m of joint in 1 hour, therefore he will point 6 m in 36 minutes.	
36 minutes bricklayer at £5.00	3.00
1 m³ of mortar will point about 2100 m of joint, therefore for 6 m of pointing the cost will be approximately	0.11
	8.11

$$\text{Cost per m} = \frac{£8.11}{6} = \underline{£1.352}$$

Unit rate = £1.35 per m (labour £1.33, materials £0.02)

Rake out joint of existing brickwork for horizontal lead flashing, and point in cement mortar 1:3 mix. Per linear metre

Owing to the work being in old brickwork with mortar that has probably set very hard, the raking-out output could fall to about 4 m per hour.

	£
Bricklayer 1 hour at £5.00	5.00
The output for pointing will fall, because of unevenness of old bricks, to about 8 m per hour. Therefore pointing 4 m will take a bricklayer half an hour at £5.00 per hour.	2.50
Mortar for pointing, say 11p	0.11
	7.61

$$\text{Cost per m} = \frac{£7.61}{4} = \underline{£1.90}$$

Unit rate = £1.90 per m (labour £1.87, materials £0.03)

Rake out horizontal joint of brickwork, enlarge for nib of asphalt skirting and point in cement mortar 1:3 mix. Per linear metre

	£
A bricklayer will rake out and enlarge about 3 m of joint in one hour.	
1 hour bricklayer at £5.00	5.000
A bricklayer will point about 9 m of joint in one hour, therefore pointing 3 m will take 20 mins.	
20 minutes bricklayer at £5.00 per hour	1.667
Mortar required, say 11p	0.110
	6.777

$$\text{Cost per m} = \frac{£6.777}{3} = \underline{£2.259}$$

Unit rate = £2.26 per m (labour £2.22, materials £0.04)

Cut vertical chase in brickwork for small pipe and make good. Per linear metre (using hand labour, small quantity assumed)

	£
1 bricklayer will cut about 2 m of chase in 1 hour.	
1 hour bricklayer at £5.00	5.00
A bricklayer will make good about 10 m of chase in 1 hour, therefore 2 m in 12 minutes.	
12 minutes bricklayer at £5.00 per hour	1.00
Mortar for making good, say 15p	0.15
	6.15

$$\text{Cost per m} = \frac{£6.15}{2} = \underline{£3.075}$$

Unit rate = £3.08 per m (labour £3.00, materials £0.08)

Cut vertical chase in brickwork for small pipe, and make good. Per linear metre (using electric power tool, large quantity assumed)

	£
Using a power tool a bricklayer will cut about 6 m of chase in 1 hour.	
1 hour bricklayer at £5.00	5.00
1 hour hire use, including power, of electric power tool at £1.50 per hour	1.50
A bricklayer will make good about 6 m in 40 minutes.	
40 minutes bricklayer at £5.00 per hour	3.33
Mortar for making good, say 15p	0.15
	9.98

$$\text{Cost per m} = \frac{£9.98}{6} = £1.66$$

Unit rate = £1.66 per m (labour £1.39, materials £0.03, plant £0.25)

Fix only tiled, slabbed fireplace 950 × 900 mm overall with slabbed hearth, fire-brick back and all-night type grate including pointing fireback with fire cement, and bedding in fireclay. Each

	£
6 firebricks at 50 p each	3.00
Fireclay, say 12 kg at 20 p per kg	2.40
Cement mortar 0.04 m³ at £39.60 per m³	1.58
Backing to interior in either common bricks or concrete, say 20 common bricks at £65.00 per thousand	1.30
It will take a bricklayer about 6 hours when assisted by a labourer full-time, to offload, distribute on site, assemble and fix this fireplace complete.	
6 hours bricklayer at £5.00 per hour	30.00
6 hours labourer at £4.00 per hour	24.00
	62.28

Unit rate = £62.28 each (labour £54.00, materials £8.28)

215 mm diameter, 300 mm high red terracotta chimney pot, set and flaunched up solid in cement mortar 1:3 mix. Each

	£
Cost of chimney pot to purchase, £4.00	4.00
0.03 m³ of cement mortar (including waste) at £39.60 per m³	1.19
This will be fixed with the general walling, therefore an attendant labourer's time should be included.	
Bricklayer fixing and flaunching ½ hour at £5.00	2.50
20 minutes attendant labourer at £4.00 per hour	1.33
	9.02

Unit rate = £9.02 each (labour £3.83, materials £5.19)

Circular fireclay flue liner 225 mm external diameter to comply with BS 1181, butt-jointed and set in flue including all bends. Per linear metre

The flue-liner is inserted during the construction of the general walling, therefore an attendant labourer's time must be included. A bricklayer will fix about 4 m of flue-liner in 1 hour. The brickwork for this liner would be measured solid without any deduction and this can offset the extra labour in forming and cutting.

	£
4 m flue-liners at £8.00 per m	32.00
Allow the extra over costs of (say) 2 bends	3.50
Waste, 5%	1.78
Mortar for surrounding, say 0.04 m³ at 39.60 per m³	1.58
Bricklayer 1 hour at £5.00	5.00
40 minutes labourer at £4.00	2.67
	46.53

$$\text{Cost per m} = \frac{£46.53}{4} = \underline{£11.63}$$

Unit rate = £11.63 per m (labour £1.92, materials £9.71)

Two courses of blue Welsh slates as horizontal damp-proof course laid to break joint bedded in cement mortar 1:3 mix. Per square metre

With 350 × 215 mm damp-proof course slates approximately 25 slates are required to cover 1 m² of wall with two courses of slate. A bricklayer will lay damp-proof course in slate to about 1 m² of wall in 1 hour.

	£
25 slates at £60.00 per 100, delivered site	15.00
Waste on slates, 10%	1.50
Mortar required for 1 m², say 0.02 m³ at £39.60 per m³	0.79
Offloading and stacking slates, 2 labourers 1 hour each to offload 1000 = £8.00	
Cost of offloading 25 = $\frac{£8.00}{1000} \times 25 =$	0.20
Bricklayer 1 hour at £5.00	5.00
Attendant labourer 30 minutes at £4.00 per hour (laid by a 2:1 gang in foundations)	2.00
	24.49

Unit rate = £24.49 per m² (labour £7.20, materials £17.29)

Asbestos-based damp-proof course to comply with BS 743, laid horizontal and bedded in cement mortar 1:3 mix (measured net). Per square metre

	£
Cost per m²	2.00
Waste, 5%	0.10
Laps, 5%	0.10
Bedding mortar is disregarded, since this will have been included with the brickwork. A bricklayer will lay 1 m² in about ¼ hour.	
¼ hour bricklayer at £5.00 per hour	1.25
⅛ hour attendant labourer (laid by a 2:1 gang in foundation work) at £4.00 per hour	0.50
	3.95

Unit rate = £3.95 per m² (labour £1.75, materials £2.20)

Fibre base bituminous felt damp-proof course to comply with BS 743, laid horizontal and bedded in cement mortar 1:3 mix (measured net). Per square metre

	£
Cost per m²	1.40
Waste, 5%	0.07
Laps, 5%	0.07
Bricklayer ¼ hour (as before)	1.25
Labourer ⅛ hour (as before)	0.50
	3.29

Unit rate = £3.29 per m² (labour £1.75, materials £1.54)

Hole for small pipe through half-brick wall. Each

It is assumed that the bricklayer will return to cut all holes, and therefore will not require the services of an attendant labourer. It will take a bricklayer about 20 minutes by hand to cut hole and make good.

	£
20 minutes bricklayer at £5.00 per hour	1.67

Unit rate = £1.67 each (labour £1.67)

Hole for small pipe through one-brick wall. Each

It will take a bricklayer about ½ hour to cut hole and make good.

	£
½ hour bricklayer at £5.00 per hour	2.50

Unit rate = £2.50 each (labour £2.50)

Hole for small pipe through two-brick wall. Each

It will take a bricklayer about 1¼ hours to cut hole and make good.

	£
1¼ hour bricklayer at £5.00 per hour	6.25

Unit rate = £6.25 each (labour £6.25)

Hole for large pipe through half-brick wall. Each

It will take a bricklayer about ½ hour to cut such a hole and make good.

	£
½ hour bricklayer at £5.00 per hour	2.50

Unit rate = £2.50 each (labour £2.50)

Hole for large pipe through one-brick wall. Each

It will take a bricklayer about 1 hour to cut hole and make good.

	£
1 hour bricklayer at £5.00 per hour	5.00

Unit rate = £5.00 each (labour £5.00)

Hole for large pipe through two-brick wall. Each

	£
The making good of such a hole will require about 4 bricks.	
4 common bricks, say 6½p each	0.26
Mortar, say 10p	0.10
It will take a bricklayer about 2 hours to cut hole and make good.	
2 hours bricklayer at £5.00 per hour	10.00
	10.36

Unit rate = £10.36 each (labour £10.00, materials £0.36)

Increase the labour outputs of the foregoing items by 33% for holes through old brick walls. The use of mechanical aids should also be considered if there are sufficient holes, as this will reduce the cost and also the total time span.

Extra over common brickwork in 65 mm bricks for facing externally in local rustic facings manufactured by Messrs. XYZ Ltd., laid English bond, pointed with a weather joint as the work proceeds. Per square metre

From the table given previously, it will be seen that a m² of facing laid English bond in 65 mm bricks requires 89 bricks.

	£
Cost of facing-bricks per thousand, delivered site	100.00
Offloading and stacking will take a labourer about 1½ hours per thousand.	
1½ hours labourer at £4.00 per hour (Alternatively the manufacturer will supply and unload the bricks in bundles on a craned lorry at an extra cost)	6.00
	106.00
Deduct cost of 65 mm common bricks	65.00
Extra cost of facings per thousand	41.00
Extra cost of 89 bricks $= \dfrac{£41.00}{1000} \times 89 =$	3.65
Waste on facings, 2½% (extra allowance)	0.09
Mortar has been included in the general walling.	
Walling facing-bricks and pointing will cause a fall in production. Allow an extra ½ hour per m² for a bricklayer only setting and pointing facings at £5.00	2.50
	6.24

Unit rate = £6.24 per m² (labour £2.50, materials £3.74)

Extra over common brickwork in 65 mm common bricks for rustic facing bricks, laid stretcher bond, and pointed with a weather joint as the work proceeds. Per square metre

From the table given previously, it will be seen that 1 m² of facing laid stretcher bond in 65 mm bricks requires 59 bricks.

	£
Extra cost of 59 facings at an extra cost of £41.00 per thousand $= \frac{£41.00}{1000} \times 59$	2.42
Waste on facings, 2½% (extra allowance)	0.06
Additional labour in laying and pointing, say, 25 minutes bricklayer at £5.00	2.08
	4.56

Unit rate = £4.56 per m² (labour £2.08, materials £2.48)

Half-brick-thick (102 mm) skin of hollow wall in 65 mm rustic facing bricks, as before, laid stretcher bond in cement mortar 1:3 mix and pointed with a weather joint as the work proceeds. Per square metre

	£
A half-brick wall in 65 mm bricks requires 59 bricks per m². 59 bricks at £106.00 per thousand, delivered and offloaded $= \frac{£106.00}{1000} \times 59$	6.25
Waste on bricks, 5%	0.31
0.03 m³ cement mortar at £39.60 per m³	1.19
When walling a half-brick wall in facing-bricks and pointing, a bricklayer's output will fall from 50 to about 40 bricks per hour. Therefore a 3:2 gang will lay 120 bricks per hour at £23.00 per gang hour.	
Cost of laying 59 $= \frac{£23.00}{120} \times 59$	11.31
	19.06

Unit rate = £19.06 per m² (labour £11.31, materials £7.75)

One-brick-thick (215 mm) wall in 65 mm rustic facing bricks, as before, laid English bond in cement mortar 1:3 mix, and weather-pointed as work proceeds. Per square metre

1 m² of one-brick-thick wall in 65 mm bricks requires 118 bricks irrespective of the bond.

	£
Cost of 118 bricks $= \dfrac{£106.00}{1000} \times 118$	12.51
Waste on bricks, 5%	0.63
0.07 m³ of cement mortar at £39.60 per m³	2.77
When walling a one-brick wall in facing-bricks pointed both sides the bricklayer's output will fall from 50 to about 35 bricks per hour. Therefore a 3:2 gang will give an hourly output of 105 bricks.	
Cost of laying 118 $= \dfrac{£23.00}{105} \times 118$	25.85
	41.76

Unit rate = £41.76 per m² (labour £25.85, materials £15.91)

Rake out joints of existing brickwork and re-point in cement mortar 1:3 mix. Per square metre

If this work is carried out in conjunction with new work, it is possible that scaffolding will be already in position. If, however, the work comprises re-pointing an existing building the scaffolding must either be included as a lump sum in the preliminaries section of the bill of quantities, or be included in the unit rate. For re-pointing old brickwork, about 0.003 m³ of mortar is required per m² which includes waste in droppings.

	£
0.003 m³ of mortar at £39.60 per m³	0.12
Bricklayer 1 hour raking out joints, per m² at £5.00	5.00
Bricklayer ½ hour pointing at £5.00	2.50
	7.62

The labour rate is dependent on the hardness of existing mortar, and the use of mechanical aids should be considered.

Unit rate = £7.62 per m² (labour £7.50, materials £0.12)

Extra over common brickwork in 65 mm common bricks for facing externally in Class A engineering bricks, laid English bond and pointed with a weather joint as the work proceeds. Per square metre

		£
Cost of Class A engineering bricks, per thousand, delivered site	*c/fwd*	£300.00

	b/fwd	£300.00
Offloading, stacking, etc. will take a labourer about 2 hours per thousand.		
2 hours at £4.00		8.00
		£308.00
Deduct cost of 65 mm commons		65.00
Extra costs of engineering bricks per thousand		243.00
$\frac{£243.00}{1000} \times 89 =$		21.63
Waste, 2½% (extra allowance)		1.08
Mortar has been included with the common brickwork.		
The labour output will drop considerably with this type of brick.		
Allow an extra hour per m² for a bricklayer only setting and pointing heavy engineering bricks		
1 hour at £5.00		5.00
		27.71

Unit rate = £27.71 per m² (labour £5.00, materials £22.71)

Extra over common brickwork in 65 mm common bricks for forming a flat arch in 65 mm rustic facing-bricks, as before, 215 mm high on face, 103 mm wide on soffit, and pointed where exposed. Per linear metre

Consider an arch 1.5 m long, which will require 20 bricks.
Extra cost of facings, at £100.00 per thousand, over commons (from previous analysis) = £41.00 per thousand, delivered and offloaded.

		£
Extra cost of 20 bricks $\frac{£41.00}{1000} \times 20$		0.82
When walling arches and pointing, a bricklayer's output will fall from 50 to about 25 bricks per hour. In previous analysis of common brickwork with an output of 50 bricks per hour, the cost of a 3:2 gang was taken as £23.00, giving an output of 150 bricks. Therefore cost included for laying 20 commons		
$\frac{£23.00}{150} \times 20 = \underline{£3.07}$		
When output falls to 25 bricks per hour, the cost of laying 75 bricks = £23.00. Cost of laying 20 bricks with this output		
	c/fwd	£0.82

b/fwd	£0.82
$= \frac{£23.00}{75} \times 20 = \underline{£6.13}$	
Therefore extra cost of laying 20 facings in arches at an output of 25 bricks per hour = £6.13 − £3.07 =	3.06
	3.88

Cost per m $= \frac{£3.88}{3} \times 2 = \underline{£2.59}$

Unit rate = £2.59 per m (labour £2.04, materials £0.55)

External sill formed with two courses of 267 × 164 mm plain roofing tiles laid to break joint, in cement mortar 1:3 mix, set weathering and projecting, including pointing and cutting. Per linear metre

	£
Consider a sill 1 m long. This will require 12 tiles. Cost of 267 × 164 mm roofing tiles = £160.00 per thousand.	
Offloading will take a labourer about 1 hour at £4.00 per hour. Cost of tiles, offloaded and stacked, £164.00.	
Cost of 12 tiles $= \frac{£164.00}{1000} \times 12$	1.97
Waste on tiles, 10%	0.20
Mortar required for bedding will be about 0.01 m^3, at £39.60 per m^3	0.40
A bricklayer will fix sills after the general walling is completed, and therefore will not require the services of an attendant labourer. It will take a bricklayer about 1 hour to construct this sill.	
1 hour bricklayer at £5.00 per hour	5.00
	7.57

Unit rate = £7.57 per m^2 (labour £5.00, materials £2.57)

76 mm thick blockwork partition wall in 448 × 219 mm solid breeze concrete blocks to comply with BS 2028 Type B in gauged mortar 1:1:6 mix. Per square metre

	£
2 bricklayers with 1 attendant labourer will wall about 4 m^2 of partition in 1 hour.	
4 m^2 at £4.80 per m^2, delivered site	19.20
c/fwd	£19.20

		£
	b/fwd	19.20
Waste on blocks, 5%		0.96
Mortar required = 0.02 m² at £34.97 per m³ (from previous analysis)		0.70
Offloading and stacking 4 m² of blocks will take a labourer about 10 minutes. Labourer 10 minutes at £4.00 per hour		0.67
Bricklayer 2 hours at £5.00 per hour		10.00
Attendant labourer 1 hour at £4.00		4.00
		35.53

Cost per m² $= \frac{£35.53}{4} = \underline{£8.88}$

Unit rate = £8.88 per m² (labour £3.67, materials £5.21)

102 mm thick blockwork partition wall in 448 × 219 mm solid breeze concrete blocks to comply with BS 2028 Type B in gauged mortar 1:1:6 mix. Per square metre

	£
2 bricklayers with 1 attendant labourer will wall about 3 m² of 102 mm partition in 1 hour. 3 m² of partition at £6.00 per m², delivered site	18.00
Waste on blocks, 5%	0.90
Mortar required = 0.02 m³ at £34.97 per m³	0.70
Offloading and stacking 3 m², say 10 minutes labourer as before (heavier blocks) at £4.00	0.67
Bricklayer 2 hours (as before) at £5.00	10.00
Labourer 1 hour (as before) at £4.00	4.00
	34.27

Cost per m² $= \frac{£34.27}{3} = \underline{£11.42}$

Unit rate = £11.42 per m² (labour £4.89, materials £6.53)

7 Rubble walling and masonry

The price of stone varies considerably, it is dependent on the type of material, size, shape and finish of the stones, location of the quarry, and particularly the haulage costs. Stone can be sold by weight or volume. In many areas it is now impossible to employ a mason and a bricklayer would carry out the work involved.

Dry stone walling average 200 mm on bed in 50 mm average courses in Yorkshire sandstone (garden walls). Per square metre

Walling average 200 mm on bed will require about ⅓ tonne of stone per square metre.

	£
⅓ tonne sawn slab ends Yorkshire stone 50 mm thick at £50.00 per tonne, delivered site and tipped	16.67
Waste on stone, 10%	1.67
It will take a mason about 2½ hours to wall 1 m² of 50 mm stone 200 mm on bed.	
2½ hours mason at £5.00 per hour	12.50
1¼ hours attendant labourer at £4.00 per hour	5.00
	35.84

Unit rate = £35.84 per m² (labour £17.50, materials £18.34)

Dry stone walling average 200 mm on bed in 50 mm average courses in second-hand Yorkshire sandstone (garden walls). Per square metre

The price of second-hand stone varies considerably depending upon the amount of demolition being carried out in the area. When demolition contractors wish to dispose at any cost, prices can be as low as £10.00 per tonne, which is merely the haulage cost. Haulage distances affect the cost of second-hand stone and can increase the cost by as much as £10.00 per tonne per additional 50 kilometres.

	£
⅓ tonne second-hand stone at £15.00 per tonne, tipped	5.00
Waste on stone, 15% (more waste than new stone)	0.75
It will take a mason about 4 hours to wall 1 m² 200 mm on bed, owing to the cutting of stones.	
4 hours mason at £5.00 per hour	20.00
2 hours labourer at £4.00 per hour	8.00
	33.75

Unit rate = £33.75 per m² (labour £28.00, materials £5.75)

Stone-walling average 200 mm on bed in 50 mm average courses, in Yorkshire sandstone in gauged mortar 1:1:6 mix, pointed on face. Per square metre

	£
⅓ tonne of stone per square metre at £50.00 per tonne	16.67
Waste on stone, 10%	1.67
A square metre 200 mm thick on bed with 50 mm deep courses will require about 0.06 m³ of mortar	
0.06 m³ of gauged mortar at £34.97	2.10
It will take a mason about 3 hours to wall in 1 m².	
3 hours mason at £5.00 per hour	15.00
1½ hours labourer at £4.00 per hour	6.00
	41.44

Unit rate = £41.44 per m² (labour £21.00, materials £20.44)

Stone-walling outer skin of hollow wall 20 mm on bed build snecked in second-hand Yorkshire sandstone in gauged mortar 1:1:6 mix, pointed on face. Per square metre

	£
⅓ tonne second-hand wallstones at £15.00 per tonne delivered site, tipped	5.00
Waste on stone, 15%	0.75
Mortar (as before)	2.10
It will take a mason about 5 hours to wall and point 1 m² of this walling.	
5 hours at £5.00 per hour	25.00
2½ hours attendant labourer at £4.00 per hour	10.00
	42.85

Unit rate = £42.85 per m² (labour £35.00, materials £7.85)

Rubbed Portland stone facework 150 mm thick, built against brick backing in lime mortar 1:3 mix, and pointed. Per square metre

	£
Visualize an area of 1 m², with stones 300 × 300 mm on face and 150 mm thick. Quotation from stone suppliers for these stones ready-rubbed and squared for walling at £150.00 per m².	150.00
Allow a nominal wastage factor of 2½% (i.e. 1 stone in 40 damaged)	3.75
Mortar required = 0.02 m³, including waste, because of the very thin joints with this type of masonry.	
0.02 m³ at, say £40.00 per m³	0.80
Offloading and stacking on site, ½ hour labourer at £4.00 per hour	2.00
Hoisting, setting and bedding in position will take a mason and a labourer working 1:1 about 2¾ hours per m².	
2¾ hours mason at £5.00 per hour	13.75
2¾ hours labourer at £4.00 per hour	11.00
To wash down stone and remove protective slurry coating, mason will take about 2 hours per m².	
2 hours mason at £5.00 per hour	10.00
	191.30

Unit rate = £191.30 per m² (labour £36.75, materials £154.55)

225 × 75 mm blue Yorkshire sandstone sill rubbed where exposed, weathered, throated, grooved and bedded in gauged mortar 1:1:6 mix. Per linear metre

	£
Consider a sill 2 metres long.	
Quotation from quarry for supply of sill ready worked to specification	70.00
Waste, nominal 2½%	1.75
Mortar for jointing and pointing, say	0.35
Offloading and stacking sill on site will take two labourers 5 minutes each. 10 minutes labourer at £4.00 per hour	0.67
It will take a mason and labourer working together 1 hour to take sill from store, hoist on the scaffold, place into position, bed and point.	
Mason 1 hour at £5.00 per hour	5.00
Labourer 1 hour at £4.00 per hour	4.00
	81.77

Cost per metre $= \frac{\pounds 81.77}{2} = \underline{\pounds 40.89}$

Unit rate = £40.89 per m (labour £4.84, materials £36.05)

8 Roofing

General contractors very seldom carry out roofing work. In recent years this work has been done by specialist sub-contractors, who offer highly competitive prices and service, because of the specialized labour that they employ and their attractive quantity discounts for materials.

508 × 254 mm blue Welsh slates, laid to a 100 mm lap, each slate double-nailed at centre with copper nails, laid on and including 38 × 25 mm softwood battens. Per square metre

£

Consider an area of 1 m²

The number of slates required can be obtained by dividing the total area by the area of the exposed portion of the slate.

$$\text{Gauge} = \frac{508 - 100}{2} = 204 \text{ mm}$$

$$\left(\text{For head-nailing, gauge} = \frac{508 - 100 - 25}{2} = 192 \text{ mm} \right)$$

Exposed portion of slate = 204 × 254 mm

$$\text{Number of slates per m}^2 = \frac{1 \text{ m}^2}{204 \times 254 \text{ mm}} = 19.3$$

19.3 slates at £140.00 per hundred, delivered site.

Offloading and stacking slates by hand will take two labourers about 10 minutes each for 100 slates.

Offloading 100 = 20 minutes at £4.00 = £1.33

19.3 slates at £141.33 per hundred

	£
$= \frac{£141.33}{100} \times 19.3$	27.28
Transit waste, purchaser's responsibility, 2½%	0.68
Fixing waste, 5%	1.36
c/fwd	£29.32

	£
b/fwd	29.32
Each slate double-nailed totals 40 nails. 38 mm copper nails at 400 nails per kg $\frac{40}{400} \times 1\ \text{kg} = 0.10\ \text{kg}$	
0.10 kg of copper nails at £5.50 per kg =	0.55
Waste on nails, 10%	0.06
The number of 1 m lengths of 38 × 25 mm batten in 1 m^2 of roof area for slates laid to a 204 mm gauge $= \frac{1000\ \text{mm}}{204\ \text{mm}} = 4.9$	
Total length of batten in 1 m^2 = 4.9 × 1 m = 4.9 m at £0.18 per m =	0.88
Waste on battens, 10%	0.09
50 mm galvanized nails for fixing battens to rafters at say 400 mm centres = say 4 nails per metre length, 5 lengths each with 4 nails = 20 nails at 280 per kg $= \frac{20}{280} \times 1\ \text{kg} = 0.07\ \text{kg}$ at 60p per kg including 10% waste	0.05
The labour output depends upon the size of the slates and the gauge. It will take a craftsman and a labourer about ¼ hour each to batten and felt 1 m^2 with battens laid to 204 mm gauge. Note that under the Standard Method of Measurement felting is measured separately, but the labour involved in battening cannot easily be separated from felting, and therefore is included here.	
¼ hour craftsman at £5.00 per hour	1.25
¼ hour labourer at £4.00 per hour	1.00
It will take a craftsman and a labourer about ⅓ hour each to lay 504 × 254 mm slates to a 204 mm gauge per m^2	
⅓ hour craftsman at £5.00 per hour	1.67
⅓ hour labourer at £4.00 per hour	1.33
	36.20

Unit rate = £36.20 per m^2 (labour £5.25, materials £30.95)

One layer of reinforced sarking felt to BS 747 Type 1F, weighing 16 kg per 15 m² roll secured to rafters with galvanized clout-headed nails, and lapped 75 mm at side and end joints. Per square metre

	£
15 × 1 m roll = 15 m²	
Cost per roll, delivered site = £11.00	
Cost per m² = $\frac{£11.00}{15}$ = £0.73 per m²	0.73
Laps, 10%	0.07
Waste, 5%	0.04
The labour of fixing the felt has been included with the battens in the slating analysis. The cost of nails has also been included in the slating.	
	0.84

Unit rate = £0.84 per m² (materials £0.84)

267 × 164 mm plain smooth faced machine-made tiles to BS 402 laid to a 67 mm lap, each tile nailed with two copper nails on 38 × 25 mm battens. Per square metre

	£
Consider an area of 1 m²	
Gauge = $\frac{267 - 67}{2}$ = 100 mm	
Area of exposed tile = 100 × 164 mm	
Number of tiles per m² = $\frac{1 \text{ m}^2}{100 \times 164 \text{ mm}}$ = 61	
Tiles cost £200.00 per thousand, delivered site.	
Offloading and stacking tiles 1 hour labourer = £4.00	
Cost of 62 tiles at £204.00 per thousand offloaded	
= $\frac{£204.00}{1000}$ × 61	12.44
Transit waste, 2½% purchaser's responsibility	0.31
Fixing waste, 5%	0.62
Nails, 62 × 2 = 124 at 400 per kg	
= $\frac{124}{400}$ × 1 kg = 0.31 kg	
0.31 kg at £5.50 per kg =	1.71
Waste on nails, 10%	0.17
c/fwd	£15.25

		£
	b/fwd	15.25
Number of 1 m lengths of 38 × 25 mm battens in 1 m² when laying tiles to a 100 mm gauge = $\frac{1\text{ m}}{100\text{ mm}} = 10$		
Total length of batten in 1 m² = 10 × 1 m = 10 m at £0.18 per m =		1.80
Waste on battens, 10%		0.18
50 mm nails for fixing battens at 400 mm centres = say 4 per metre length × 10 = 40 nails at 280 per kg $\frac{40}{280} \times 1$ kg = 0.14 kg at 60p per kg =		0.08
Waste on nails, 10%		0.01
Lathing and felting for tiles laid to 100 mm gauge will take a craftsman and a labourer about ½ hour each per m²		
½ hour craftsman at £5.00 per hour		2.50
½ hour labourer at £4.00 per hour		2.00
It will take a craftsman and a labourer about ¾ hour each to lay double nailed tiles to an area of 1 m² at 100 mm gauge.		
¾ hour craftsman at £5.00		3.75
¾ hour labourer at £4.00 per hour		3.00
		28.57

Unit rate = £28.57 per m² (labour £11.25, materials £17.32)

380 × 280 mm red precast concrete, single lap interlocking standard tiles to comply with BS 550, laid to a 75 mm lap, each tile every fourth course double-nailed with aluminium nails, laid on 38 × 25 mm battens. Per square metre

		£
Consider an area of 1 m²		
A 380 × 230 mm tile with a 75 mm single lap and 30 mm side laps will have an exposed area of 305 × 200 mm.		
Number of tiles required per m² = $\frac{1\text{ m}^2}{305 \times 200} = 16.5$		
Cost of tiles £280.00 per thousand delivered site.		
Offloading and stacking tiles will take a labourer about 1½ hours at £4.00 per hour = £6.00		
Cost of 1000 tiles offloaded and stacked = £286.00		
Cost of 16.5 tiles = $\frac{16.5}{1000} \times 286.00$		4.72
	c/fwd	£4.72

		£
	b/fwd	4.72
Transit waste, 2½% purchaser's responsibility		0.12
Laying waste, 5%		0.24
Nailing every fourth course will mean that only 25% of tiles nailed (4 tiles). Each tile double nailed = 8 nails at 320 per kg = 0.025 kg at £4.60 per kg		0.12
Waste on nails, 10%		0.01
Number of 1 m battens in 1 m² when laying tiles to a 305 mm gauge = 3⅓		
Total length of batten in 1 m² = 3⅓ × 1 m = 3⅓ m at £0.18 per m = £0.60		0.60
Waste on battens, 10%		0.06
50 mm nails for fixing battens at 400 mm centres = say 4 nails per metre length × 3⅓ lengths = 14 nails at 280 per kg.		
$\frac{14}{280}$ × 1 kg = 0.05 kg at 60 p per kg including 10% waste		0.03
Lathing and felting for tiles to a 305 mm gauge will take a craftsman and a labourer about ⅙ hour.		
⅙ hour craftsman at £5.00 per hour		0.83
⅙ hour labourer at £4.00 per hour		0.67
It will take a craftsman and a labourer about ⅕ of an hour each to lay 380 mm single lap tiles to a 305 mm gauge nailed every fourth course.		
⅕ hour craftsman at £5.00 per hour		1.00
⅕ hour labourer at £4.00 per hour		0.80
		9.20

Unit rate = £9.20 per m² (labour £3.30, materials £5.90)

Extra over for 250 mm diameter half-round ridge tiles to match general roofing bedded and pointed in tinted mortar. Per linear metre

	£
Half-round ridge tiles 500 mm long.	
2 tiles at £2.00 each	4.00
Waste, 5%	0.20
Tinted mortar, allow 0.01 m³ at, say £40.00 per m³	0.40
A craftsman and a labourer will bed and point about 10 metres per hour.	
⅒ hour craftsman at £5.00 per hour	0.50
⅒ hour labourer at £4.00 per hour	0.40
	5.50

Unit rate = £5.50 per m (labour £0.90, materials £4.60)

Raking cutting on 380 × 230 mm interlocking pantiles. Per linear metre

As the main item of tiling is measured net, the cutting at rake will involve an additional waste allowance which depends on the roof pitch. Allow 4 extra tiles per linear metre of cutting.

	£
4 tiles at £286.00 per thousand	1.14
A craftsman will measure and cut 4 tiles in 10 minutes.	
⅙ hour craftsman at £5.00 per hour	0.83
	1.97

Unit rate = £1.97 per m (labour £0.83, materials £1.14)

No. 5 milled lead in covering to flat roof. Per square metre

No. 5 lead weighs 24 kg/m².

	£
1 tonne milled lead, delivered site, at £1000.00 per tonne	1000.00
Waste on lead, 2½% (very low because of good scrap value for waste)	25.00
According to the Standard Method of Measurement, lead in flat roofs is billed in square metres. This area includes several allowances for labours such as drips, rolls and upstands to walls. In fact, the area of flat roof covered by the lead area given in the bill of quantities is only about two-thirds. If laid flat, 1 tonne of No. 5 lead would cover about 41 m² of roof, but the actual roof area covered and visible would be about 27 m². It will take a plumber and mate about 2 hours each to lay 1 m² of flat roof and 82 hours for 1 tonne including unloading.	
Plumber 82 hours at £5.00 per hour	410.00
Plumber's mate 82 hours at £4.00 per hour (a plumber's mate is classed as a labourer for cost purposes, but in practice is an apprentice in the last few years of his training)	328.00
Cost per tonne laid	1763.00

1 tonne of No. 5 lead, when laid flat, will cover about 41 m² of roof area. As all allowances have been made in the area given in the bill of quantities, the covering capacity of 41 m² can be used here.

$$\text{Cost per m}^2 = \frac{£1763.00}{41} = \underline{£43.00}$$

Unit rate = £43.00 per m² (labour £18.00, materials £25.00)

No. 4 milled lead cover-flashing 175 mm girth, lapped 100 mm at intermediate joints with lead tacks at 1 m centres, turned into brick joint and secured with lead wedges (measured net). Per linear metre

	£
No. 4 lead weighs 19 kg/m².	
1 tonne of No.4 lead, delivered site	1000.00
Waste, 2½%	25.00
As the quantity given in the bill of quantities is net, an allowance must be made for laps, tacks and lead wedges. This amounts to about 20%	205.00
1 tonne of No. 4 lead would cover about 51 m² or about 291 m of 175 mm girth flashing. It will take a plumber and mate about 20 minutes each to fix 1 metre of flashing, say, 100 hours per tonne including unloading.	
100 hours plumber at £5.00 per hour	500.00
100 hours plumber's mate at £4.00 per hour	400.00
	2130.00

1 tonne of lead, with 20% added for laps, passings, tacks and wedges, will cover 291 m of 175 mm girth flashing.

$$\text{Cost per linear metre} = \frac{£2130.00}{291} = £7.32$$

Unit rate = £7.32 per m (labour £3.09, materials £4.23)

No. 4 milled lead in stepped flashing, average 200 mm girth, lapped 100 mm at intermediate joints with lead tacks at 0.5 m centres, turned into brick joints and secured with lead wedges. Per linear metre

	£
1 tonne of No. 4 milled lead, delivered site	1000.00
Waste, 2½%	25.00
Laps, passings, tacks and wedges, 20%	205.00
1 tonne of No. 4 lead will cover about 51 m² or 255 m of 200 mm girth flashing. It will take a plumber and mate about 30 minutes to fix 1 m of stepped flashing, say, 130 hours per tonne including unloading.	
130 hours plumber at £5.00 per hour	650.00
130 hours plumber's mate at £4.00 per hour	520.00
	2400.00

$$\text{Cost per linear metre} = \frac{£2400.00}{255} = £9.41$$

Unit rate = £9.41 per m (labour £4.59, materials £4.82)

9 Woodwork

Buying timber

Timber for carpentry and joinery can be purchased from timber merchants in a sawn state, planed, or planed and moulded. The basic selling unit is the cubic metre but joinery rails, etc., are normally quoted per linear metre. The price paid by a builder varies according to the amount of timber purchased. If sawn sections and lengths are required which correspond exactly to the members which arrive from the timber-exporting countries, the price will be cheaper than if a builder requires specific lengths and sections which need cutting. The emphasis in recent years has been to prefabricate joinery components in a specialist sub-contractor's workshop wherever possible and reduce the site labour to a minimum. Examples of such items are roof trusses, doors and frames, windows, staircases, cupboard units, etc. It is now very unusual to find any of these items manufactured on a building site. Some *in situ* work is still necessary in respect of certain carcassing members, flooring, rails, fascias, shelving and special fittings. Joinery members would normally be purchased from a merchant with all the labours and mouldings machined before delivery to site. Provided a particular labour is standard and a reasonable quantity of timber is ordered, the cost of such machining is very small. If specific lengths of, say, floor joists are delivered to site, the cost will be higher than for random lengths but there will be an obvious saving of waste because cutting is unnecessary.

Waste

The cutting and wastage factor on *in situ* carpentry and joinery can be set generally at about 10 per cent but for certain members in some situations can rise to a far higher figure. It is considered necessary to allow a nominal 2½ per cent wastage on all prefabricated joinery components. In the case of doors, for example, it is reasonable to assume that one door in forty will be damaged, lost or stolen. It has been argued that the contractor is insured for such eventualities but in practice he cannot make frequent claims for these relatively minor losses. He must therefore cover himself by adding a nominal sum to the cost of each prefabricated component. This principle has been extended throughout the book to all similar items such as ironmongery, sanitary appliances, etc. The increases in mechanical unloading, vandalism and theft have highlighted this particular problem.

Formula for calculating the cost of timber

For timber prices quoted in metres cube, the cost per linear metre for any scantling size can be calculated from the following: multiply the cross-sectional area of the timber in square millimetres (mm^2) by the price in pounds and move the decimal point six places to the left. The result is the price of the timber in decimals of a pound per linear metre for that scantling size. It is of course necessary to obtain an accurate price per cubic metre by considering such factors as quantity and the actual sections and mouldings, etc., required.

Hardwood

The labour rates shown in this chapter refer generally to working with softwood. If hardwood is specified it will be necessary to increase the labour cost according to the hardness of the particular material. As an approximate general rule, 50 per cent should be added to softwood rates for working with hardwood.

Generally

Pricing measured joinery work can be very complicated in practice especially with certain hardwoods. An estimator working for a specialist joinery manufacturer would have a completely different approach to an estimator from a general contractor who did not have his own specialist joinery shop.

100 × 75 mm softwood plate. Per linear metre

	£
1 m^2 of carcassing softwood costs £125.00 delivered site.	
Offloading and stacking will take a labourer about 1 hour at £4.00 per hour.	
Therefore cost of timber unloaded and stacked =	129.00
Waste on timber in general construction work, 10%	12.90
Nails required for general construction work is about 1 kg per m^3 at 60p per kg	0.60
Waste on nails, 10%	0.06
It will take a joiner about 12 hours to fix 1 m^3 of timber in wall plates.	
12 hours at £5.00 per hour	60.00
Cost per m^3 =	202.56

$$\text{Cost per metre} = £202.46 \times \frac{100 \times 75}{1000 \times 1000} = \underline{£1.520}$$

Unit rate = £1.52 per m (labour £0.45, materials £1.07)

150 × 50 mm sawn softwood carcassing in floors. Per linear metre

	£
1 m³ of carcassing softwood (as before), offloaded and stacked on site	129.00
Waste, 10%	12.90
Nails (as before)	0.60
Waste on nails, 10%	0.06
It will take a joiner an average of about 20 hours to cut to length, set out and fix 1 m³ of timber in floor carcassing members, but this time will increase for members having small cross-section dimensions.	
20 hours at £5.00 per hour	100.00
Cost per m³ =	242.56

$$\text{Cost per metre} = £242.56 \times \frac{150 \times 50}{1000 \times 1000} = \underline{£1.820}$$

Unit rate = £1.82 per m (labour £0.75, materials £1.07)

100 × 50 mm sawn softwood carcassing in pitched roofs. Per linear metre

	£
1 m³ of carcassing timber (as before)	129.00
Waste, 10%	12.90
Nails (as before)	0.60
Waste on nails, 10%	0.06
It will take a joiner an average of about 30 hours to cut to length, set out, notch and fix 1 m³ of timber in pitched roof carcassing but this time will increase for members having small cross-section dimensions.	
30 hours at £5.00 per hour	150.00
Cost per m³ =	292.56

$$\text{Cost per metre} = £292.56 \times \frac{100 \times 50}{1000 \times 1000} = \underline{£1.463}$$

Unit rate = £1.46 per m (labour £0.75, materials £0.71)

Note: A contractor may purchase carcassing members cut to specific lengths in which case he will be charged about 10 per cent more for the timber but the site waste factor can then be reduced to about 5 per cent. The labour rate can also be reduced by at least 10 per cent in respect of site cutting to lengths.

25 mm thick herringbone strutting between 225 mm deep joists. Per linear metre

This item is measured over the joists and is billed in linear metres. A joiner will fix about 3 m of herringbone strutting in 1 hour. If the joists are at 400 mm centres, the strutting between two joists will have two 500 mm lengths, i.e. 1 m of 50 × 25 mm timber.

Quantity of material required for 3 m $= \dfrac{3\text{ m}}{400\text{ mm}} \times 1\text{ m}$

$= \dfrac{3000}{400} \times 1\text{ m} = 7.5\text{ m}$ of 50 × 25 mm timber

	£
Cost of 1 metre of 50 × 25 mm softwood (i.e. 0.00125 m^3) at £129.00 per m^3 = £0.161 per metre.	
7.5 m at £0.161	1.21
Waste, say, 10%	0.12
Nails: each strut double nailed at each end will require about 56 No. 50 mm nails at 350 per kg = 0.143 kg at 60p per kg, say	0.09
Waste on nails, 10%	0.01
Joiner 1 hour at £5.00 per hour	5.00
	6.43

Cost per metre $= \dfrac{£6.43}{3} = \underline{£2.14}$

Unit rate = £2.14 per m (labour £1.67, materials £0.47)

150 × 25 mm wrot softwood fascia board. Per linear metre

A joiner will fix about 6 m of fascia board in 1 hour.

	£
Cost of 6 m of 150 × 25 mm wrot softwood fascia board at £1.30 per m	7.80
Waste, 10%	0.78
63 mm nails allow two every 400 mm for nailing to rafters = 30 nails for 6 metres at 220 per kg, say	0.06
Waste on nails, 10%	0.01
Joiner 1 hour at £5.00 per hour	5.00
	13.65

Cost per linear metre $= \dfrac{£13.65}{6} = \underline{£2.28}$

Unit rate = £2.28 per m (labour £0.83, materials £1.45)

300 mm wide soffit in 6 mm flat asbestos cement sheeting. Per linear metre

	£
A joiner will cut and fix about 4 m of 300 mm wide asbestos soffit board in 1 hour. 4 m × 300 mm = 1.2 m².	
1.2 m² of 6 mm flat asbestos cement sheeting at £1.50 per m²	1.80
Waste on asbestos sheeting 20%	0.36
Joiner 1 hour at £5.00 per hour	5.00
	7.16

$$\text{Cost per metre} = \frac{£7.16}{4} = £1.79$$

Unit rate = £1.79 per m (labour £1.25, materials £0.54)

Floorboarding

Floorboarding is marketed in either linear metres or square metres. As the width of a tongue is approximately 12 mm the timber lost in the tongues is inversely proportional to the board width, for example, for 50 mm wide boards, tongue = 25%, and for 100 mm wide boards, tongue = 12½%. Unless the boarding is sold by area, on the assumption that 1 m² supplied will cover 1 m² of floor area, it will be necessary to make an allowance for the area 'lost' by the tongues.

25 mm nominal thickness wrot softwood tongued and grooved strip flooring in 75 mm face widths double-nailed to each joist with 63 mm floor brads, cramped up, nails punched and puttied and dressed off smooth and level. Per square metre

	£
1 m² of flooring in 75 mm widths	4.50
Unloading, etc. Labourer 100 m² per hour at £4.00 $= \frac{£4.00}{100} =$	0.04
Add an allowance for tongues if not included in the quoted price per m² $\frac{12\text{ mm tongue}}{75\text{ mm width}} = 16\%$	0.73
	5.27
Waste, 10%	0.53
Assume joists at 400 mm centres and each board double nailed. Allow ½ kg at 60p per kg	0.30
c/fwd	£6.10

	£
b/fwd	6.10
Waste on nails, 10%	0.03
It will take a joiner about ¾ hour to lay 1 m² of floorboarding in 75 mm widths.	
Joiner ¾ hour at £5.00 per hour	3.75
	9.88

Unit rate = £9.88 per m² (labour £3.75, materials £6.13)

25 mm nominal thickness wrot softwood tongued and grooved board flooring in 150 mm face widths double-nailed to each joist with 63 mm floor brads, cramped up, nails punched and puttied and dressed off smooth and level. Per square metre

	£
1 m² of flooring in 150 mm widths	4.20
Unloading (as before)	0.04
Add tongue allowance as before $\frac{12 \text{ mm}}{150 \text{ mm}} = 8\%$	0.34
	4.58
Waste 10%	0.46
Nails, allow ¼ kg at 60p per kg	0.15
Waste 10%	0.02
It will take a joiner about 35 minutes to lay 1 m² of floorboarding in 150 mm widths.	
Joiner 35 minutes at £5.00 per hour	2.92
	8.13

Unit rate = £8.13 per m² (labour £2.92, materials £5.21)

100 × 19 mm wrot softwood skirting, moulded and plugged to brick walls, including all ends, mitres and angles. Per linear metre

	£
Cost of 10 m of wrot softwood skirting ready-moulded from merchants at £0.50 per metre	5.00
Waste, 15% (very heavy waste is incurred with this type of timber)	0.75
Timber for plugs, nil (waste used)	
Skirting double-nailed at say 1 m centres = 10 × 2 = 20, 63 mm oval brad-head nails at 240 per kg = 1/12 kg at 60p per kg, say	0.05
Waste on nails, 10%	0.01
c/fwd	£5.81

		£
	b/fwd	5.81
If the skirting is nailed every 1 m, 10 plugs will be required, and it will take a joiner about 10 minutes to make each plug, say 1⅔ hours altogether.		
Joiner 1⅔ hours plugging at £5.00 per hour		8.33
It will take a joiner about 1 hour to cut and fix 10 m of softwood skirting including all labours.		
Joiner 1 hour fixing at £5.00 per hour		5.00
		19.14

Cost per linear metre $= \frac{£19.14}{10} = \underline{£1.91}$

Unit rate = £1.91 per m (labour £1.33, materials £0.58)

38 × 25 mm sawn softwood open spaced grounds plugged to brick walls, fixed at 500 mm centres. Per square metre

	£
Consider an area of 1 square metre.	
Grounds fixed at 500 mm centres in all directions will require 4½ metres of ground per m²	
4.5 m of 38 × 25 mm softwood grounds at £0.20 per metre	0.90
Waste, 10%	0.09
Each ground double-nailed to plugs at 500 mm centres will require 6 plugs and 12 nails, 63 mm round-wire nails at 275 kg = $\frac{63}{275}$ = say ¼ kg at 60p per kg	0.15
Waste on nails, 10%	0.02
It will take a joiner about 1 hour to form 6 plugs and nail the grounds.	
Joiner 1 hour at £5.00 per hour	5.00
	6.16

Unit rate = £6.16 per m² (labour £5.00, materials £1.16)

Woodworking machinery

When joinery is shop-manufactured by machine, many of the labours can be done in one operation. The following examples show how machine labours can be costed, and are based upon a four-cutter electrical woodworking machine at a cost of, say, £6.00 per hour including operative. Visualize 1000 metres of 75 × 50 mm softwood frame, rebated, throated, grooved, and moulded – four labours in all.

	£
Setting up machine, grinding and shaping cutters, man and machine 4 hours at £6.00 per hour.	24.00
Machining timber at 8 metres per minute, 2 hours 5 minutes at £6.00 per hour	12.50
Labourers offloading timber in machine shop, and stacking and re-stacking 3.75 m³ at 3 hours per m³ at £4.00 per hour	45.00
	81.50

Cost per linear metre $= \dfrac{£81.50}{1000} = \underline{£0.08}$

Cost per labour (four labours on timber) $= \dfrac{£0.08}{4}$
$= £0.02$

Visualize 100 metres of timber of the same section to be machined.

	£
Setting up machine as before, 4 hours at £6.00 per hour	24.00
Machining timber 8 metres per minute, 12½ minutes at £6.00 per hour	1.25
Labourers handling 0.375 m³ of timber at 3 hours per m³ at £4.00 per hour	4.50
	29.75

Cost per linear metre $= \dfrac{£29.75}{100} = \underline{£0.30}$

Cost per labour (four labours on timber)
$= \dfrac{£0.30}{4} = £0.08$

It can be seen from the analysis above that the quantity of work to be machined considerably affects the price of each labour. An estimator working for a builder who does not have a specialist joinery shop will have to obtain quotations which include all planing and machining costs. In the case of most items this will result in the basic cost per m³ of sawn joinery timber (purchased in large quantities) being at least doubled in respect of machined and moulded joinery members.

100 × 75 mm door frame in wrot softwood, moulded, rebated and plugged to brick walls. Per linear metre

Consider a frame for a door size 826 × 2040 mm which, including horns, requires about 5.2 m of frame.

	£
100 × 75 mm wrot joinery timber costs £2.20 per metre including unloading costs.	
5.2 m of 100 × 75 mm frame at £2.20 per m	11.44
Two machine labours on 5.2 m of timber at, say, £0.05 per labour per m	0.52
	11.96
Waste on machined frame, 15%	1.79
12 No. 150 mm nails for site fixing at 36 per kg = ⅓ kg at 60 p per kg	0.20
Waste on nails, 10%	0.02
Man and machine about ¼ hour cutting out frame and mortising and tenoning at £6.00 per hour	1.50
Joiner knocking up frame and jointing about ¼ hour at £5.00 per hour	1.25
Joiner fixing on site about ½ hour at £5.00 per hour	2.50
	19.22

Cost per linear metre $= \dfrac{£19.22}{5.2} = \underline{£3.70}$

Unit rate = £3.70 per m (labour £0.91, materials £2.59, plant £0.20)

150 × 38 mm wrot softwood door lining, rebated, plugged to brickwork, and tongued at angles. Per linear metre

	£
Consider a lining for a door size 826 × 2040 mm which requires about 5 m of lining.	
150 × 38 mm wrought joinery timber costs £1.80 per m including unloading costs.	
5 m of 150 × 38 mm lining at £1.80 per m =	9.00
Rebating by machine labour 5 m at £0.05 per m	0.25
Waste on machined lining, 15%	1.39
12 No. 100 mm nails for site fixing at 66 per kg = say 1/6 kg at 60p per kg	0.10
Waste on nails, 10%	0.01
Man and machine about ¼ hour cutting out lining and tongueing and grooving at £6.00 per hour	1.50
Joiner knocking up lining on site and fixing in opening, say ¾ hour at £5.00 per hour	3.75
	16.00

Cost per linear metre $= \dfrac{£16.00}{5} = \underline{£3.20}$

Unit rate = £3.20 per m (labour £0.95, materials £2.10 plant £0.15)

The last two examples illustrate the costs when a builder has his own joinery shop. Different calculations are necessary if a builder purchases prefabricated frames or linings, as illustrated in the next example.

100 × 75 mm door frame as before described. Per linear metre

	£
Consider a frame for a door size 826 × 2040 mm which, including horns, has a measurement in the bill of quantities of about 5.2 m of frame	
Cost of one frame delivered pre assembled with protective batten, say	14.50
Nominal waste, 2½%	0.36
Nails including waste as before	0.22
Labourer offloading, etc. 5 minutes at £4.00 per hour	0.33
Joiner fixing ½ hour at £5.00 per hour	2.50
	17.91

Cost per linear metre $= \dfrac{£17.91}{5.2} = £3.44$

Unit rate = £3.44 per m (labour £0.54, materials £2.90)

Note: It is also possible to enumerate door frames, etc. in accordance with SMM clause N.17. The last example would then read:

Unit rate = £17.91 each (labour £2.83, materials £15.08)

75 × 19 mm wrot softwood architraves, splayed, including all ends, mitres, and angles. Per linear metre

	£
Consider a set of architraves for a door 826 × 2040 mm. Such a door will require about 5.35 m of architrave.	
75 × 19 mm wrought joinery timber costs £0.45 per metre including unloading costs.	
5.35 m of 75 × 19 mm architraves at £0.45 per m	2.41
Waste, 15%	0.36
12 No. 50 mm nails at 350 per kg = 1/30 kg at 60p per kg including waste	0.02
It will take a joiner about ½ hour to cut to length and fix this set of architraves.	
½ hour joiner at £5.00 per hour	2.50
	5.29

Cost per linear metre $= \dfrac{£5.29}{5.35} = £0.99$

Unit rate = £0.99 per m (labour £0.47, materials £0.52)

44 mm thick standard flush door size 826 × 2040 mm with solid core, covered both sides with birch-faced plywood, and hardwood lipped at edges, for painting. Each

The SMM states that standard ready-made doors are to be enumerated, and that hanging is deemed to be included.

	£
Cost of door to purchase in lots of 20	21.00
Nominal waste, 2½%	0.53
Joiner hanging door on butt hinges, about 1 hour.	
1 hour joiner at £5.00 per hour (note that hinges are measured separately)	5.00
	26.53

Unit rate = £26.53 each (labour £5.00, materials £21.53)

There are many standard types and sizes of doors available at varying prices. If specific doors are required which are not stock pattern, these would normally be ordered from a door manufacturer. It is possible that a builder who owned his own joinery shop might make these doors himself but in either case, unless mass production was possible, the costs would be very high.

19 × 13 mm wrot softwood rounded glazing beads. Per linear metre

	£
Consider a pane 250 × 250 mm with pinned beads.	
1 m of bead ready-rounded at £0.30 per m	0.30
Waste, 20%	0.06
Pins including waste, say	0.03
Joiner cutting to length, mitring and temporarily fitting, say 10 minutes at £5.00 per hour	0.83
	1.22

Unit rate = £1.22 per m (labour £0.83, materials £0.39)

19 × 13 mm wrot oak rounded glazing beads secured with brass cups and screws. Per linear metre

	£
Consider a pane 250 × 250 mm with screwed beads.	
1 m of bead ready-rounded at £0.50 per m	0.50
Waste, allow 33⅓%	0.17
12 brass screws at £3.00 per 100	0.36
12 brass cups at £0.75 per 100	0.09
Waste on cups and screws, 10%	0.04
Joiner cutting to length, mitring and temporarily fitting, say ¼ hour at £5.00 per hour	1.25
	2.41

Unit rate = £2.41 per m (labour £1.25, materials £1.16)

Standard ready-made softwood window to comply with BS 644 Part I, type 240 V, overall size 1225 × 1225 mm, complete with easy-clean hinges and aluminium alloy ironmongery to all opening lights. Each

	£
Cost of ready-made window delivered to site, all as specification from joinery manufacturer	28.00
Nominal waste, 2½%	0.70
Offloading and stacking, say 2 labourers 2½ minutes each = 5 minutes labourer at £4.00 per hour	0.33
Preparing and fixing in opening	
Joiner ¾ hour at £5.00 per hour	3.75
	32.78

Unit rate = £32.78 each (labour £4.08, materials £28.70)

If the window is build in as the brickwork is being carried out there may be an overlapping of the duties of the joiner and bricklayer.

19 mm thick birch-faced blockboard shelving 200 mm wide. Per linear metre

A joiner will cut, prepare and fix about 4 m of shelving in 1 hour.

	£
1 m of 19 mm blockboard 200 mm wide at £5.00 per m²	1.00
Waste, 20%	0.20
Nails including waste, say	0.10
Joiner ¼ hour at £5.00 per hour	1.25
	2.55

Unit rate = £2.55 per m (labour £1.25, materials £1.30)

50 × 38 mm sawn softwood framed bearers. Per linear metre

	£
Consider a frame 1 m × 1 m with one intermediate rail for a purpose-made fitting.	
5 m of 50 × 38 mm softwood at £0.60 per metre	3.00
Waste, 20%	0.60
Nails (including waste) say	0.06
Framing with half-lap joints by hand labour will take a joiner about ½ hour.	
½ hour joiner at £5.00 per hour	2.50
	6.16

$$\text{Cost per m} = \frac{£6.16}{5} = \underline{£1.23}$$

Unit rate = £1.23 per m (labour £0.50, materials £0.73)

Supply and fix a pair of 100 mm cast steel butt hinges to softwood. Per pair

	£
Cost of 1 pair of 100 mm cast steel butt hinges	0.70
Nominal waste, 2½%	0.02
Screws, 16 required at £1.50 per 100	0.24
Waste on screws, 10%	0.02
Labour of fixing included with the hanging of doors elsewhere	
	0.98

Unit rate = £0.98 each (materials £0.98)

75 mm three-lever mortice dead lock fixed to softwood. Each

	£
Cost to purchase, complete with screws, say	3.50
Nominal waste, 2½%	0.09
Joiner fixing 1½ hours at £5.00 per hour	7.50
	11.09

Unit rate = £11.09 each (labour £7.50, materials £3.59)

75 mm three-lever mortice lock fixed to softwood. Each

	£
Cost to purchase, complete with screws, say	4.00
Nominal waste, 2½%	0.10
Joiner fixing 1½ hours at £5.00 per hour	7.50
	11.60

Unit rate = £11.60 each (labour £7.50, materials £4.10)

Cylinder rim latch fixed to softwood. Each

	£
Cost to purchase including screws and skeleton keys, say	4.80
Nominal waste, 2½%	0.12
Joiner fixing 1½ hours at £5.00 per hour	7.50
	12.42

Unit rate = £12.42 each (labour £7.50, materials £4.92)

75 mm three-lever rebated mortice lock fixed to softwood. Each

	£
Cost to purchase, say	5.50
Nominal waste, 2½%	0.14
Joiner fixing 1¾ hours at £5.00 per hour	8.75
	14.39

Unit rate = £14.39 each (labour £8.75, materials £5.64)

150 × 15 mm BMA flush bolt with receiver for shoot and fix to softwood. Each

	£
Cost to purchase, including BMA screws, say	2.50
Nominal waste, 2½%	0.06
Joiner fixing ¾ hour at £5.00 hour	3.75
	6.31

Unit rate = £6.31 each (labour £3.75, materials £2.56)

Set of silver anodized aluminium lever latch door furniture, spring-loaded and fixing to softwood. Per set

	£
Cost to purchase, including matching screws, say	2.50
Nominal waste, 2½%	0.06
Joiner fixing ¼ hour at £5.00 per hour	1.25
	3.81

Unit rate = £3.81 each (labour £1.25, material £2.56)

Set of silver anodized aluminium lever lock door furniture, spring-loaded, and fixing to softwood. Per set

	£
Cost to purchase, including matching screws, say	2.80
Nominal waste, 2½%	0.07
Joiner fixing ¼ hour at £5.00 per hour	1.25
	4.12

Unit rate = £4.12 per set (labour £1.25, material £2.87)

300 × 75 mm BMA finger plate, fixing to softwood. Each

	£
Cost to purchase, including BMA screws, say	3.20
Nominal waste, 2½%	0.08
Joiner fixing 10 minutes at £5.00 per hour	0.83
	4.11

Unit rate = £4.11 each (labour £0.83, materials £3.28)

Standard model pneumatic overhead door-closer with 90-degree retention, sprayed bronze finish, and fixing to softwood. Each

	£
Cost to purchase, including matching screws, say	25.00
Nominal waste, 2½%	0.63
Joiner fixing and adjusting 3 hours at £5.00 per hour	15.00
	40.63

Unit rate = £40.63 each (labour £15.00, materials £25.63)

Double-action floor spring with 90-degree retention and BMA finish where exposed, including adjustable top centre and loose box for concrete. Each

	£
Cost of 1 spring to purchase, say	65.00
Nominal waste 2½%	1.63
Cement and sand for grouting, say 50p	0.50
2 joiners, 3 hours each hanging heavy hardwood door on floor spring and top pivots (the mortice in the floor will have been previously measured).	
Joiner 6 hours at £5.00 per hour	30.00
	97.13

Unit rate = £97.13 each (labour £30.00, materials £67.13)

250 mm BMA letter plate fixed to softwood. Each

	£
Cost to purchase, say	5.50
Nominal waste, 2½%	0.14
Joiner fixing 1½ hours at £5.00 per hour	7.50
	13.14

Unit rate = £13.14 each (labour £7.50, materials £5.64)

BMA hat and coat hook fixed to softwood. Each

	£
Cost to purchase, including BMA screws, say	0.75
Nominal waste, 2½%	0.02
Joiner fixing 5 minutes at £5.00 per hour	0.83
	1.60

Unit rate = £1.60 each (labour £0.83, materials £0.77)

30 mm diameter BMA concave screw cupboard door-knob. Each

	£
Cost to purchase	1.20
Nominal waste, 2½%	0.03
Joiner fixing 10 minutes at £5.00 per hour	0.83
	2.06

Unit rate = £2.06 each (labour £0.83, materials £1.23)

250 × 200 mm black japanned pressed-steel shelf brackets fixed to softwood. Each

	£
Cost to purchase, including screws	0.35
Nominal waste, 2½%	0.01
Joiner fixing 10 minutes at £5.00 per hour	0.83
	1.19

Unit rate = £1.19 each (labour £0.83, materials £0.36)

Note: For fixing ironmongery to hardwood the labour constants should be increased by at least 25 per cent.

10 Plasterwork and other floor, wall and ceiling finishes and beds

Lime plaster has a backing or rendering of coarse stuff and a finishing or setting coat of fine stuff. Before the price of lime plastering can be analysed, the basic cost of coarse stuff and fine stuff must be ascertained.

The labour in mixing plastering materials is covered by the time of the attendant labourers in the plastering gang, as two labourers can supply three plasterers with all the materials they require and do all the mixing. Generally the mixing is carried out by hand and not by mixing machines.

Coarse stuff (1:2:9 mix for render backings for lime plaster)

Coarse Sirapite plaster including unloading, etc. (1 hour labourer per tonne) £58.00 + £4.00 = £62.00 per tonne.

	£
1 m³ at 1000 kg/m³ at £62.00	62.00
Hydrated lime including unloading, etc. £88.00 + £4.00 = £92.00 per tonne.	
2 m³at 700 kg/m³at £92.00	128.80
Plastering sand	
9 m² at 1600 kg/m³ at £10.00 per tonne	144.00
	334.80
Allow for shrinkage, consolidation and waste, add 50% (a factor of 50% should be included because of the wastage of the constituent materials at all stages)	167.40
	502.20

Divide by parts of mix (i.e. 12) $\frac{£502.20}{12}$

Cost per cubic metre = £41.85

Labour mixing is included in the plastering gang. The use of animal hair is now unlikely. An alternative mix would be to use ordinary Portland cement instead of plaster for the rendering coat.

Fine stuff (1:4:4 mix for setting coats in lime plaster)

	£
Hydrated lime including unloading, etc.	
1 m³ at 700 kg/m³ at £92.00	64.40
Fine Sirapite plaster including unloading, etc.	
£56.00 + £4.00 = £60.00	
4 m³ at 1000 kg/m³ at £60.00 per tonne	240.00
Plastering sand.	
4 m³ at 1600 kg/m³ at £10.00 per tonne	64.00
	368.40
Allow for shrinkage, consolidation and waste, add 50%	184.20
	552.60

Divide by parts of mix (i.e. 9) $\frac{£552.60}{9}$

Cost per cubic metre =	£61.40

Labour mixing is included in the plastering gang.

Render and set (two-coat work) 15 mm finished thickness in Sirapite lime sand coarse stuff, 1:2:9 mix, and Sirapite lime sand fine stuff, 1:4:4 mix, steel float finish, to brick and block walls. Per square metre

	£
12 mm rendering coat at £41.85 per cubic metre	
$= \frac{12}{1000} \times £41.85$	0.50
3 mm setting or skimming coat at £61.40 per cubic metre	
$= \frac{3}{1000} \times 61.40$	0.18
A plasterer with full attendance will apply 1 m² of rough render in about 10 minutes, and 1 m² of setting coat in about 10 minutes. A plastering gang of 3 plasterers to 2 labourers will be used; therefore allow ⅔ of the plasterer's time for the attendant labourers.	
Say, 21 minutes plasterer at £5.00 per hour	1.75
14 minutes attendant labourer at £4.00 per hour	0.93
	3.36

Unit rate = £3.36 per m² (labour £2.68, materials £0.68)

Render and set to brick walls in lime plaster as before, in widths not exceeding 300 mm. Per square metre

	£
Materials cost for 1 m²	0.68
Labour cost of 1 m² on general walling is £2.68	
Work in narrow widths will cause a fall in output, therefore the labour cost will be increased by 50% = £2.68 + 50%	4.02
	4.70

Unit rate = £4.70 per m² (labour £4.02), materials £0.68)

Render and set (two-coat work) 12 mm finished thickness in gypsum hardwall plaster, having fibred browning undercoat 1:2½ mix, neat gypsum plaster finishing coat, and steel float finish, to brick and block walls. Per square metre

	£
Rendering coat material.	
Browning backing including unloading etc., £56.00 + £4.00 = £60.00 per tonne.	
1 m³ at 1000 kg/m³ at £60.00 per tonne	60.00
Plastering sand	
2½ m³ at 1600 kg/m³ at £10.00 per tonne	40.00
	100.00
Shrinkage, consolidation and waste, add 50%	50.00
	150.00
Divide by parts of mix (i.e. 3½) $\frac{£150.00}{3½}$	
Cost per cubic metre =	£42.86

Setting coat material	
Gypsum finishing including unloading, etc. £54.50 + £4.00 = £58.50 per tonne.	
1 m³at 1000 kg/m³ at £58.50 per tonne	58.50
Shrinkage, consolidation and waste, add 50%	29.25
Cost per cubic metre =	£87.75

10 mm rendering coat at £42.86 per cubic metre $= \frac{10}{1000} \times £42.86$	0.43
2 mm setting coat at £87.75 per cubic metre $= \frac{2}{1000} \times £87.75$	0.18
c/fwd	£0.61

		£
	b/fwd	0.61
Hardwall plaster is more difficult than lime plaster to apply as the material sets rapidly, therefore allow 15 minutes per square metre for each coat.		
30 minutes plasterer at £5.00 per hour		2.50
20 minutes labourer at £4.00 per hour		1.33
		4.44

Unit rate = £4.44 per m² (labour £3.83, materials £0.61)

Render and set to brick walls in gypsum plaster, as before, but in compartments not exceeding 4 m² on plan. Per square metre

	£
Materials cost for 1 m²	0.61
Labour cost for 1 m² general walling = £3.83	
Work in confined areas in compartments not exceeding 4 m² on plan will cause a fall in output, therefore the labour cost will be increased by 25% = £3.83 plus 25%	4.79
	5.40

Unit rate = £5.40 per m² (labour £4.79, materials £0.61)

Note: Generally all finishings to compartments, staircase areas, etc., should have the labour costs increased by about 25 per cent because of lower outputs from the operatives due to limited working space and the need to move the men, materials and plant more often.

Render and set (two-coat work) 10 mm finished thickness in Carlite pre-mixed plaster, in accordance with the manufacturer's instructions, steel-float finish, to concrete ceilings. Per square metre

Carlite bonding coat.

1 tonne of pre-mixed bagged plaster at £82.00 per tonne plus 1 hour labourer offloading at £4.00 = £86.00

1 tonne of pre-mixed Carlite bonding plaster will cover about 160 m² if spread 8 mm thick, including waste and allowance for compression (from manufacturer's supplied information).

Therefore cost of material required for 1 m²

$$= \frac{£86.00}{160} = £0.54$$

Carlite setting coat.
1 tonne of pre-mixed bagged plaster at £65.00 per tonne plus 1 hour labourer offloading = £69.00
1 tonne of pre-mixed Carlite finishing plaster will cover about 450 m² if spread 2 mm thick, including waste and allowance for compression (from manufacturer's supplied information).

	£
Therefore cost of material required for 1 m² $= \frac{£69.00}{450} = £0.15$	
Bonding plaster for 1 m²	0.54
Finishing plaster for 1m²	0.15
Working with a quick-setting material and overhead to ceilings, a plasterer's output will fall, therefore allow 18 minutes per coat per m². Owing to the plaster being pre-mixed, the labour gang can be reduced to 2 plasterers and 1 attendant labourer.	
36 minutes plasterer at £5.00 per hour	3.00
18 minutes labourer at £4.00 per hour	1.20
	4.89

Unit rate = £4.89 per m² (labour £4.20, materials £0.69)

Render, float and set (three-coat work) 15 mm finished thickness in gypsum plaster, having fibred browning rendering and floating coats 1:2½ mix, and neat gypsum plaster finishing coat, steel-float finish, on expanded metal lathing ceilings. Per square metre

	£
Material for 8 mm thick rendering coat at £42.86 per cubic metre (as previously calculated) $= \frac{8}{1000} \times £42.86 =$	0.34
Material for 5mm thick floating coat at £42.86 per m³ $= \frac{5}{1000} \times £42.86$	0.21
Material for 2 mm thick setting coat at £87.75 per m³ as previously calculated $= \frac{2}{1000} \times £87.75 =$	0.18
c/fwd	£0.73

The render coat is comparatively easy to apply: it will take a plasterer about 10 minutes for 1 m². The floating and setting coats will be applied normally, as before, therefore allow 15 minutes per coat.

		£
	b/fwd	0.73
Render coat	10 minutes	
Floating coat	15 minutes	
Setting coat	15 minutes	
	40 minutes	
40 minutes plasterer at £5.00 per hour		3.33
Working a 3:2 gang, the labourer's time required will be approximately 27 minutes at £4.00 per hour		1.80
		5.86

Unit rate = £5.86 per m² (labour £5.13, materials £0.73)

No. 26 gauge (0.56 mm) expanded metal lathing secured to softwood ceilings with galvanized staples. Per square metre

	£
Consider an area of 10 m²	
10 m² of expanded metal at £1.60 per m²	16.00
Cutting waste on sheet material, 10%	0.16
An area of 10 m² will require about 1 kg of staples at 85p per kg	0.85
Waste on staples, 10%	0.09
Offloading and distributing 10 m² of metal about the site will take 2 labourers about 5 minutes each.	
10 minutes labourer at £4.00 per hour	0.67
2 plasterers with 1 attendant labourer will fix 10 m² of metal to ceilings in about 1½ hours.	
3 hours plasterer at £5.00 per hour	15.00
1½ hours labourer at £4.00 per hour	6.00
	38.77

Cost per square metre $= \frac{£38.77}{10} = £3.88$

Unit rate = £3.88 per m² (labour £2.17, materials £1.71)

9.5 mm thick gypsum plaster baseboard to comply with BS 1230, secured to softwood joists with 12 mm galvanized clout-headed nails, joints covered with jute scrim cloth. Per square metre

Consider an area of 10 m² using 914 × 1200 mm sheets
to joists at 450 mm centres

	£
10 m² of plasterboard at £1.12 per m² =	11.20
Offloading and stacking will take a labourer about 5 minutes per 10 m² at £4.00 per hour	0.33
Cutting waste on sheet material, 10%	1.15
Nailing 914 mm wide sheets to joists at 450 mm centres requires 3 rows of nails per sheet = 39 nails.	
For 10 m² approximately 9 sheets are required = 9 × 39 nails = 351 at 500 per kg = 0.70 kg at 90p per kg =	0.63
Waste on nails, 10%	0.06
An average of 20 m of scrim cloth will be required per 10 m² of boards at £2.40 per 100 metre roll	0.48
Waste on scrim cloth, 10%	0.05
1 plasterer attended by 1 labourer will board and scrim 10 m² in about 1½ hours.	
1½ hours plasterer at £5.00	7.50
1½ hours attendant labourer at £4.00 per hour	6.00
	27.40

Cost per square metre = $\frac{£27.40}{10}$ = £2.74

Unit rate = £2.74 per m² (labour £1.35, materials £1.39)

Fine plaster solid cornice, filleted and curved, 400 mm girth on face. Per linear metre

A plasterer with an attendant labourer will run about 2 m of 400 mm girth cornice in one hour. About 0.06 m³ of fine stuff is required for 2 m of cornice.

	£
0.06 m³ of fine stuff gauged plaster at £66.00 per m³ (from previous analysis)	3.96
Waste due to droppings, 20%	0.79
1 hour plasterer at £5.00 per hour	5.00
1 hour labourer at £4.00 per hour	4.00
	13.75

Cost per linear metre = $\frac{£13.75}{2}$ = £6.88

Unit rate = £6.88 per m (labour £4.50, materials £2.38)

External angle to cornice. Each

It will take a plasterer about ¼ hour to form an external angle.

	£
¼ hour plasterer at £5.00 per hour	1.25
¼ hour attendant labourer at £4.00 per hour	1.00
	2.25

Unit rate = £2.25 each (labour £2.25)

Internal angle to cornice. Each

Internal angles are far more difficult to form: one will take a plasterer about ¾ hour.

	£
¾ hour plasterer at £5.00 per hour	3.75
¾ hour attendant labourer at £4.00 per hour	3.00
	6.75

Unit rate = £6.75 each (labour £6.75)

Gyproc fibrous plaster cornice 225 mm girth, fixed to walls. Per linear metre

A plasterer with an attendant labourer will fix about 5 m of such a cornice in 1 hour.

	£
5 m of cornice at £1.40 per m	7.00
Cutting waste, 10%	0.07
Neat plaster for fixing, say 30p	0.30
Plasterer 1 hour at £5.00 per hour	5.00
Labourer 1 hour at £4.00 per hour	4.00
	16.37

Cost per linear metre $= \frac{£16.37}{5} = £3.27$

Unit rate = £3.27 per m (labour £1.80, materials £1.47)

Galvanised perforated metal angle bead with 50 mm returns to comply with BS 1246, secured to brick arris with plaster dabs, including working plaster to bead both sides. Per linear metre

Generally room heights are about 2.3 m, and a plasterer will fix about five lengths to walls in one hour.

	£
11.5 m of metal angle bead at £0.40 per m	4.60
Wate, 10%	0.46
c/fwd	£5.06

		£
	b/fwd	5.06
Neat plaster for fixing, say 30p		0.30
As these beads are fixed during the general plastering working with a 3:2 gang, ⅔ hour must be included for an attendant labourer.		
1 hour plasterer at £5.00 per hour		5.00
40 minutes labourer at £4.00 per hour		2.67
		13.03

$$\text{Cost per linear metre} = \frac{£13.03}{11.5} = £1.13$$

Unit rate = £1.13 per m (labour £0.67, materials £0.46)

Cement and sand 1:3 mix in 25 mm thick screeded bed. Per square metre

The cost of the materials in this item has already been calculated at £39.60 in the brickwork chapter (first example on page 78). This amount included a 33⅓% allowance for shrinkage, wastage, etc., which would be appropriate for floor screeds as the wastage in mortar droppings would be approximately cancelled out by the extra materials necessary to allow for undulations beneath the bed.

	£
The cost of 1 m² of screeded bed = $\frac{25}{1000} \times £39.60$	0.99
(The above cost includes the use of a 5/3½ mixer)	
2 plasterers will be supplied with material by 1 labourer, who will mix the material in the mixer. One spreader will lay per hour about 4 m² with a screeded finish.	
15 minutes plasterer at £5.00 per hour	1.25
7½ minutes labourer at £4.00 per hour	0.50
	2.74

Unit rate = £2.74 per m² (labour £1.75, materials £0.99)

Cement and sand 1:3 mix in 25 mm thick trowelled bed. Per square metre

	£
Materials (as before)	0.99
Laying a 25 mm thick bed, trowelled finish, a plasterer will lay only about 3 m² per hour.	
20 minutes plasterer at £5.00 per hour	1.67
10 minutes labourer at £4.00 per hour	0.67
	3.33

Unit rate = £3.33 per m² (labour £2.34, materials £0.99)

Cement and sand 1:3 mix in 36 mm average thickness screeded bed laid to falls. Per square metre

	£
1 m² of 36 mm bed $= \frac{36}{1000} \times$ £39.60 =	1.43
A plasterer will lay only about 2 m² per hour, 36 mm thick to falls.	
30 minutes plasterer at £5.00 per hour	2.50
15 minutes attendant labourer at £4.00 per hour	1.00
	4.93

Unit rate = £4.93 per m² (labour £3.50, materials £1.43)

25 mm thick granolithic concrete floor finishing consisting of 1 part cement to 2½ parts granite chippings, 6 mm to dust laid on concrete, steel-trowelled finish. Per square metre

	£
Ordinary Portland cement at £59.00 per tonne offloaded (from previous analysis)	
1 m³ at 1420 kg/m³ at £59.00 per tonne	83.78
Granite chippings, 6 mm to dust at £20.00 per tonne, delivered site.	
2½ m³ at 1350 kg/m³ at £20.00 per tonne	67.50
	151.28
Shrinkage, consolidation and wastage *add* 33⅓%	50.43
	201.71
Divide by parts of mix (i.e. 3½)	
Cost of 1 m³ $= \frac{£201.71}{3\frac{1}{2}} =$	57.63
Use of 5/3½ mixer for one hour with an output of 1 m³ per hour (from previous analysis)	0.47
Labour of mixing and transporting on site will be included in the labour gang for laying.	
Cost per cubic metre =	£58.10

Cost of materials and mixer of 25 mm bed $= \frac{25}{1000} \times$ £58.10 =	1.45
A plasterer will lay about 2½ m² per hour.	
24 minutes plasterer at £5.00 per hour	2.00
12 minutes attendant labourer at £4.00 per hour	0.80
	4.25

Unit rate = £4.25 per m² (labour £2.80, materials £1.45)

25 mm thick granolithic concrete paving as before, but to 300 mm wide treads of staircases. Per linear metre

	£
Materials cost for one m²	1.45
A plasterer will lay about 1 m² of 25 mm thick grano to treads and risers of stairs in one hour.	
1 hour plasterer at £5.00 per hour	5.00
½ hour attendant labourer at £4.00 per hour	2.00
	8.45

$$\text{Cost per linear metre 300 mm wide} = \frac{£8.45 \times 300}{1000} = £2.54$$

Unit rate = £2.54 per m (labour £2.10, materials £0.44)

152 × 152 × 13 mm thick red quarry tile paving in cement mortar, 1:3 mix on screeded bed (measured separately). Per square metre

	£
Consider an area of 1 m²	
Cost of 1 m² of tiles, delivered site	8.50
Labour offloading and stacking, 3 minutes at £4.00 per hour	0.20
	8.75
Waste on tiles, 5%	0.44
0.01 m³ of mortar 1:3 mix at £39.60 per m³	0.40
It will take a tiler about 1¼ hours to bed and lay 1 m² of quarry tiles	
1¼ hour tiler at £5.00 per hour	6.25
⅝ hour attendant labourer at £4.00 per hour	2.50
	18.34

Unit rate = £18.34 per m² (labour £8.75, materials £9.59)

152 × 152 × 13 mm thick red quarry tile paving, bedded in cement mortar 1:3 mix, with 3 mm wide joints pointed in white cement mortar on screeded bed (measured separately). Per square metre

	£
1 m² of tiles laid (as previous analysis)	18.34
Coloured cement for pointing 1m² of paving including waste, say	0.20
Additional labour pointing 1 m², tiler only, ¾ hour at £5.00 per hour	3.75
	22.29

Unit rate = £22.29 per m² (labour £12.50, materials £9.79)

50 mm thick precast vibrated concrete paving slabs 900 × 600 mm bedded with cement mortar dabs 1:3 mix, joints filled solid, pointed, laid on bed of ashes (measured separately). Per square metre

	£
Cost of slabs per m², delivered site	3.50
Offloading and stacking on site will take 2 labourers about 2 minutes each m².	
4 minutes labourer at £4.00 per hour	0.27
	3.77
Waste on slabs, 5%	0.19
Cement mortar 1:3 mix for bedding in domino dabs and in pointing (including waste) allow	0.15
10 m² paving in about 3 hours.	
18 minutes bricklayer at £5.00 per hour	1.50
18 minutes attendant labourer at £4.00 per hour	1.20
	6.81

Unit rate = £6.81 per m² (labour £2.70, materials £4.11)

Cement and sand 1:3 mix in 12 mm thick screeded backing, applied to brick and block walls. Per square metre

	£
1 m² of screeded backing = $\frac{12}{1000}$ × £39.60 (from previous analysis)	0.48
Additional wastage of 10% for extra droppings and undulations in the background surface compared with laying in horizontal beds	0.05
2 plasterers attended by 1 labourer will each lay about 5 m² of backing in about 1 hour.	
12 minutes plasterer at £5.00 per hour	1.00
6 minutes attendant labourer at £4.00 per hour	0.40
	1.93

Unit rate = £1.93 per m² (labour £1.40, materials £0.53)

152 × 152 × 6 mm thick white glazed wall tiles, bedded in cement mortar 1:3 mix, joints pointed in white cement, laid on screeded backing. Per square metre

	£
Cost of 1 m² of tiles, delivered site	7.50
c/fwd	£7.50

		£
	b/fwd	7.50
Labour offloading and stacking 2 minutes at £4.00 per hour		0.13
		7.63
Waste on tiles, 5%		0.38
0.01 m³ of cement mortar for bedding including waste (from previous analysis) at £39.60 per m³		0.40
White cement for slurry pointing, say 8p		0.08
A wall-tiler will work without an attendant labourer and the output will depend upon the amount of cutting involved, but on average a tiler will take about 2 hours to fix one m².		
2 hours tiler at £5.00 per hour		10.00
		18.49

Unit rate = £18.49 per m² (labour £10.00, materials £8.49)

Extra for edge tile. Per linear metre

	£
Cost of 1 m² of rounded-edge tiles	9.00
Deduct cost of 1 m² of plain tiles	7.50
Extra cost per square metre =	£1.50

Extra cost per m² = £1.50

Therefore extra cost per linear metre 152 mm wide

$$= \frac{£1.50}{1000} \times 152 = £0.23$$

There is no additional labour involved.

Unit rate = £0.23 per m (labour nil, materials £0.23)

3 mm thick hardboard plain sheet finishing to walls, pinned to grounds, to comply with BS 1142, with flush butt joints. Per square metre

		£
Consider an area of 10 m²		
Cost of 10 m² delivered site		9.00
Offloading, say, 5 minutes labourer at £4.00 per hour		0.33
Waste on sheet materials, 20%		1.87
The quantity of hardboard pins depends upon the size of sheet to be used, but on average about 300 pins will be		
	c/fwd	£11.20

		£
	b/fwd	11.20
required to fix 10 m^2. Hardboard pins at 750 per kg $= \frac{300}{750} = 0.4$ kg at 90 p per kg		0.36
Waste on pins, 10%		0.04
The fixing time depends to a large extent on the sizes of sheets used, but on average it will take two joiners about 1½ hours each to fix 10 square metres.		
3 hours joiner at £5.00 per hour		15.00
		26.60

$$\text{Cost per square metre} = \frac{£26.60}{10} = £2.66$$

Unit rate = £2.66 per m^2 (labour £1.50, materials £1.16)

11 Plumbing installations

A plumber always works with a mate, who is usually an apprentice. For estimating purposes the fact that this apprentice could be paid a proportion of a craftsman's rate is disregarded, and the mate is always charged as a labourer. Plumbers have three different grades of skill: technical, advanced and trained. A rate of £6.00 per hour has been included for a tradesman in the following examples.

100 mm diameter cast-iron half-round eaves gutter to comply with BS 460, secured to fascia with screwed brackets at 900 mm centres, jointed with compound and gutter bolts. Per linear metre

Assume gutter supplied in 1.83 m lengths.
A plumber and mate will fix 2 No. 1.83 m lengths of cast-iron gutter in 1 hour.

	£
2 lengths of 100 mm diameter cast-iron at £5.50 each	11.00
Waste in cutting to length, 5%	0.55
4 fascia brackets at 55p each	2.20
Waste on brackets, 10%	0.22
8 No. 40 mm screws for brackets at £1.50 per 100	0.12
Waste on screws, 10%	0.01
Jointing compound for two joints at say 6p per joint (including waste)	0.12
2 gutter bolts at 5p	0.10
Waste on gutter bolts, 10%	0.01
Plumber fixing 1 hour at £6.00 per hour	6.00
Mate 1 hour at £4.00 per hour	4.00
	24.33

$$\text{Cost per linear metre} = \frac{£24.33}{3.66} = \underline{£6.65}$$

Unit rate = £6.65 per m (labour £2.73, materials £3.92)

Extra for stop end on 100 mm diameter cast-iron eaves gutter. Each

	£
Cost of stop end to purchase	0.80
Waste on fittings due to loss and breakages, 5%	0.04
Displacement of gutter — nil	
Jointing compound for 1 joint including waste	0.06
Gutter bolt and waste	0.06
Plumber fixing 10 minutes at £6.00 per hour	1.00
Mate fixing 10 minutes at £4.00 per hour	0.67
	2.63

Unit rate = £2.63 each (labour £1.67, materials £0.96)

Extra for angle on 100 mm diameter cast-iron half-round eaves gutter. Each

	£
Cost of angle to purchase	2.20
Waste, 5%	0.11
	2.31
Deduct gutter displacement say 0.45 m at £5.50 per 1.83 m length plus 5% waste	1.42
Extra cost	0.89
Jointing compound for 2 joints as before	0.12
2 gutter bolts as before	0.11
Plumber fixing 15 minutes at £6.00 per hour	1.50
Mate fixing 15 minutes at £4.00 per hour	1.00
	3.62

Unit rate = £3.62 each (labour £2.50, materials £1.12)

100 mm diameter asbestos cement half-round eaves gutter, secured to fascia with screwed brackets at 900 mm centres, jointed with compound, and gutter bolts. Per linear metre

Assume gutter supplied in 1.83 m lengths.
A plumber and mate will fix about 2 No. 1.83 m lengths of gutter in 1 hour.

	£
2 lengths of 100 mm diameter asbestos gutter at £2.50 per length	5.00
Waste in cutting to length, 5%	0.25
4 fascia brackets at 40p each	1.60
Waste on brackets, 10%	0.16
c/fwd	£7.01

		£
	b/fwd	7.01
8 No. 40 mm screws for brackets at £1.50 per 100		0.12
Waste on screws, 10%		0.01
Jointing compound for 2 joints at say 6p per joint (including waste)		0.12
2 gutter bolts at 5p each		0.10
Waste on gutter bolts, 10%		0.01
Plumber fixing 1 hour at £6.00 per hour		6.00
Mate fixing 1 hour at £4.00 per hour		4.00
		17.37

Cost per linear metre $= \frac{£17.37}{3.66} = \underline{£4.75}$

Unit rate = £4.75 per m (labour £2.73, materials £2.02)

100 mm diameter PVC half-round eaves gutter, with snap joints secured to fascia with screwed brackets at 1 m centres. Per linear metre

Assume gutter supplied in 4 m lengths.
A plumber and mate will fix 2 No. 4 m lengths of PVC gutter in 1 hour.

	£
8 m of PVC gutter at £5.20 per 4 m length	10.40
Waste in cutting to length, 5%	0.52
2 fascia joint brackets at 55p each	1.10
6 fascia support brackets at 30p each	1.80
Waste on brackets, 10% of £2.90	0.29
No jointing compound required because joints snap together.	
16 No. 40 mm screws for fixing brackets at £1.50 per 100	0.24
Waste on screws, 10%	0.02
No gutter bolts required.	
Plumber fixing 1 hour at £6.00 per hour	6.00
Mate fixing 1 hour at £4.00 per hour	4.00
	24.37

Cost per linear metre $= \frac{£24.37}{8} = \underline{£3.05}$

Unit rate = £3.05 per m (labour £1.25, materials £1.80)

64 mm diameter medium grade cast-iron rainwater pipe to comply with BS 460 with spigot and socket joints with ears cast on, secured to brick walls with plastic distance pieces and hardwood plugs, and galvanized mushroom-headed pipe nails, jointed in compound. Per linear metre

Assume rainwater pipe supplied in 1.83 m lengths..
A plumber and mate will fix 2 No. 1.83 lengths of pipe in 1 hour.

	£
2 No. 1.83 m lengths of 64 mm cast-iron pipe at £12.00 per length	24.00
Waste cutting to length, 5%	1.20
Jointing compound for 2 joints at 15p per joint (including waste)	0.30
4 plastic distance pieces at 5p each	0.20
Waste on distance pieces, 10%	0.02
Timber for plugs – nil	
4 No. 75 mm galvanised mushroom-headed pipe nails at 3p each	0.12
Waste on nails, 10%	0.01
Plumber 1 hour at £6.00 per hour	6.00
Mate 1 hour at £4.00 per hour	4.00
	35.85

$$\text{Cost per linear metre} = \frac{£35.85}{3.66} = \underline{£9.80}$$

Unit rate = £9.80 per m (labour £2.73, materials £7.07)

Extra for anti-splash shoe on 64 mm diameter rainwater pipe. Each

	£
Cost of shoe to purchase	2.76
Waste, 5%	0.14
	2.90
Deduct rainwater pipe displacement of 150 mm at £12.00 per 1.83 m length plus 5% waste	1.03
Extra cost	1.87
Jointing compound for 1 joint as before	0.15
2 plastic distance pieces as before (including waste)	0.11
Timber for plugs – nil	
2 No. 75 mm galvanized nails as before (including waste)	0.07
Plumber ¼ hour fixing at £6.00 per hour	1.50
Mate ¼ hour fixing at £4.00 per hour	1.00
	4.70

Unit rate = £4.70 each (labour £2.50, materials £2.20)

Extra for 150 mm offset bend on 64 mm diameter rainwater pipe. Each

		£
Cost of bend to purchase		4.00
Waste, 5%		0.20
		4.20
Deduct rainwater pipe displacement of 450 mm at £12.00 per 1.83 m length plus 5% waste		3.10
	Extra cost	1.10
Jointing compound for two joints as before		0.30
No other materials are required for fixing an offset.		
Plumber ¼ hour fixing at £6.00 per hour		1.50
Mate fixing ¼ hour at £4.00 per hour		1.00
		3.90

Unit rate = £3.90 each (labour £2.50, materials £1.40)

Cast-iron rainwater head 250 × 175 × 175 mm, with outlet for 64 mm diameter pipe. Each

	£
Cost to purchase	5.80
Waste, 5%	0.29
Rainwater heads are measured 'full value', therefore there is no displacement of pipe to deduct.	
Jointing compound for 1 joint as before (including waste)	0.15
2 plastic distance pieces as before (including waste)	0.11
2 No. 75 mm galvanized mushroom-headed nails as before (including waste)	0.07
Plumber fixing ¼ hour at £6.00 per hour	1.50
Mate fixing ¼ hour at £4.00 per hour	1.00
	8.92

Unit rate = £8.92 each (labour £2.50, materials £6.42)

64 mm diameter asbestos cement rainwater pipe with spigot and socket joints secured to brick walls with plugs and screw-on holderbats, jointed in compound. Per linear metre

Assume rainwater pipe supplied in 1.83 m lengths.
A plumber and mate will fix 2 No. 1.83 m lengths of pipe in 1 hour.

	£
2 No. 1.83 m lengths of 64 mm asbestos pipe at £3.00 per length	6.00
Waste in cutting to length, 5%	0.30
Jointing compounds for 2 joints at 15p per joint (including waste)	0.30
2 holderbats for 64 mm pipes at 75p each	1.50
Waste on holderbats, 10%	0.15
Plumber fixing 1 hour at £6.00 per hour	6.00
Mate fixing 1 hour at £4.00 per hour	4.00
	18.25

Cost per linear metre $= \frac{£18.25}{3.66} = \underline{£4.99}$

Unit rate = £4.99 per m (labour £2.73, materials £2.26)

64 mm diameter PVC rainwater pipe secured to brick walls with special sockets, plastic distance pieces, and pipe nails. Per linear metre

Assume rainwater pipes supplied in 4 m lengths.
A plumber and mate will fix 2 No. 4 m lengths of pipe in 1 hour.

	£
2 No. 4 m lengths of pipe at £5.40 per length	10.80
Waste cutting to length, 5%	0.54
4 plastic distance pieces as before	0.20
Waste on distance pieces, 10%	0.02
Timber for plugs – nil	
4 No. 75 mm galvanized mushroom-headed pipe nails as before	0.12
Waste on nails, 10%	0.01
2 special socket connectors at £0.70 each	1.40
Waste on socket connectors, 10%	0.14
2 fixing brackets at £0.50 each	1.00
Waste on brackets, 10%	0.10
Plumber fixing 1 hour at £6.00 per hour	6.00
Mate fixing 1 hour at £4.00 per hour	4.00
	24.33

Cost per linear metre $= \frac{£24.33}{8} = \underline{£3.04}$

Unit rate = £3.04 per m (labour £1.25, materials £1.79)

101.6 mm (nominal) diameter, 4.8 mm thick cast-iron soil pipe to comply with BS 416, with spigot and socket joints and ears cast on, secured to brick walls with plastic distance pieces and hardwood plugs and mushroom-headed pipe nails, jointed with caulked lead. Per linear metre

Assume soil pipes supplied in 1.83 m lengths.
A plumber and mate will fix 1 No. 1.83 m length of pipe and joint in 1 hour.

	£
1.83 m length of 101.6 mm diameter cast-iron soil pipe at £18.00 per length	18.00
Waste cutting to length, 5%	0.90
1 kg of lead for caulking joint at £1.10 per kg (including waste)	1.10
0.2 kg of tarred gaskin for joint at £1.20 per kg (including waste)	0.24
2 plastic distance pieces at 5p each	0.10
Waste on distance pieces, 10%	0.01
Timber for plugs – nil	
2 No. 75 mm galvanized mushroom-headed pipe nails at 3p each	0.06
Waste on nails, 10%	0.01
Plumber 1 hour at £6.00 per hour	6.00
Mate 1 hour at £4.00 per hour	4.00
	30.42

$$\text{Cost per linear metre} = \frac{£30.42}{1.83} = \underline{£16.62}$$

Unit rate = £16.62 per m (labour £5.46, materials £11.16)

Extra for single-branch junction on 101.6 mm soil pipe. Each

	£
Cost of junction to purchase	11.80
Waste, 5%	0.59
	12.39
Deduct soil pipe displacement of 450 mm at £18.00 per 1.83 m length plus 5% waste	4.64
	7.75
2 kg of lead for caulking two extra joints at £1.10 per kg (including waste)	2.20
0.4 kg of tarred gaskin for caulking two extra joints at £1.20 per kg (including waste)	0.48
No other material is required for fixing a junction.	
Plumber 1 hour fixing and jointing at £6.00 per hour	6.00
Mate 1 hour at £4.00 per hour	4.00
	20.43

Unit rate = £20.43 each (labour £10.00, materials £10.43)

15 mm bore galvanized wrought-iron tubing to comply with BS 1387 Class B, with screwed joints laid in trenches. Per linear metre

Screwed wrought-iron tubing is usually marketed in 4 m to 7 m lengths and can be obtained either as plain tubes or with sockets attached. There are three grades of tube used, each grade distinguished by a coloured band painted on the pipe. For general building purposes plain tubes are used, the ends being threaded with a hand machine on site in conjunction with tapped malleable fittings. Tubes and fittings are priced on a basic list with varying percentage adjustments for quality of tubes and the size of the load.

	£
Consider 30 m of pipe in 5 m lengths.	
30 m of 15 mm diameter tube at £81.00 per 100 m	24.30
Waste on tube-cutting to lengths, 5%	1.22
30 m of tube in 5 m lengths will require 6 plain sockets at 20p each	1.20
Waste on sockets, 5%	0.06
Jointing paste, say 6p per joint (including waste) × 12	0.72
It will take a plumber about ¼ hour to thread the end of a tube and joint to the socket, therefore 6 lengths require 12 screwed ends at ¼ hour each.	
3 hours plumber at £6.00 per hour	18.00
3 hours mate at £4.00 per hour	12.00
The labour of placing the tube in the trench is included in the above time.	
	57.50

$$\text{Cost per linear metre} = \frac{£57.50}{30} = £1.92$$

Unit rate = £1.92 per m (labour £1.00, materials £0.92)

15 mm bore galvanized wrought-iron tubing to comply with BS 1387 Class B, with screwed joints, secured to brick walls with screw-on pipe brackets and plugs. Per linear metre

	£
Consider 30 m of pipe.	
30 m of 15 mm tube (as before)	24.30
Waste 5% (as before)	1.22
c/fwd	£25.52

		£
	b/fwd	25.52
Generally lengths of tubing fixed to walls are divided up into short lengths by fittings, and these are measured separately, therefore sockets will not be measured.		
Pipe brackets at say 1 m centres, 30 brackets required at 30p each		9.00
Waste on brackets, 5%		0.45
60 screws for fixing brackets at £1.50 per 100		0.90
Waste on screws, 10%		0.09
60 rawlplugs for screws at £1.00 per 100		0.60
Waste on plugs, 10%		0.06
Labour outputs vary considerably. 30 m of tube in one continuous length could be fixed with an output of 6 m per hour but as most plumbing installations have tubing divided up into short lengths because of fittings, an average output is about 3 m per hour.		
10 hours plumber at £6.00 per hour		60.00
10 hours mate at £4.00 per hour		40.00
		136.62

Cost per linear metre $= \frac{£136.62}{30} = \underline{£4.55}$

Unit rate = £4.55 per m (labour £3.33, materials £1.22)

Extra for 15 mm bore elbow. Each

	£
Cost of 1 elbow 50p	0.50
Waste on elbows, 5%	0.03
Paste for making 2 joints at 6p each	0.12
Plumber threading end of tube and jointing ¼ hour per end, ½ hour at £6.00 per hour	3.00
Mate ½ hour at £4.00 per hour	2.00
	5.65

Unit rate = £5.65 each (labour £5.00, materials £0.65)

Extra for 15 mm bore tee. Each

		£
Cost of tee to purchase 60p		0.60
Waste on tee, 5%		0.03
Paste for making 3 joints at 6p each		0.18
	c/fwd	£0.81

		£
	b/fwd	0.81
Plumber threading end of tube and jointing ¼ hour per end, ¾ hour at £6.00 per hour		4.50
Mate ¾ hour at £4.00 per hour		3.00
		8.31

Unit rate = £8.31 each (labour £7.50, materials £0.81)

Note: For elbows, tees, etc., the amount of pipe displaced is negligible and can therefore be disregarded.

13 mm (nominal) bore polythene tubing to BS 1972 (Class C) with compression-type joints laid in trench as water service pipe. Per linear metre

	£
Consider 30 m of tubing.	
Polythene tubing is marketed in 15, 30 or 60 m coils.	
2 No. 15 m coils at £4.00 per coil	8.00
Waste on tube, 5%	0.40
1 straight connector at 40p each	0.40
Waste on connector, 5%	0.02
Plumber offloading, placing in trench and making compression joint on connector, say ½ hour at £6.00 per hour	3.00
Mate ½ hour at £4.00 per hour	2.00
	13.82

Cost per linear metre $= \frac{£13.82}{30} = \underline{£0.46}$

Unit rate = £0.46 per m (labour £0.17, materials £0.29)

13 mm (nominal) bore polythene tubing to comply with BS 1972 (Class C), with compression-type joints fixed to timber with screwed brackets, as water service pipe. Per linear metre

		£
Consider 30 m of tubing.		
30 m of polythene tube (as before)		8.00
Waste on tube, 5%		0.40
	c/fwd	£8.40

	£
b/fwd	8.40
Connectors will not be required because the tubing is divided into lengths to accommodate intervening fittings.	
Polythene tubing requires bracket supports every 400 mm, therefore 75 brackets required at 15p each	11.25
Waste on brackets, 5%	0.56
150 screws for fixing brackets at £1.50 per 100	2.25
Waste on screws, 10%	0.23
It will take a plumber about 8 hours to fix 30 m of polythene tube to softwood, including screwing brackets.	
8 hours plumber at £6.00 per hour	48.00
8 hours mate at £4.00 per hour	32.00
	102.69

$$\text{Cost per linear metre} = \frac{£102.69}{30} = \underline{£3.42}$$

Unit rate = £3.42 per m (labour £2.67, materials £0.75)

Extra for 13 mm (nominal) bore elbow. Each

	£
13 mm bore polythene compression-type elbow	0.50
Waste on elbow, 5%	0.03
Plumber fixing 10 minutes at £6.00 per hour	1.00
Mate fixing 10 minutes at £4.00 per hour	0.67
	2.20

Unit rate = £2.20 each (labour £0.53, materials £1.67)

Extra for 13 mm (nominal) bore tee. Each

	£
13 mm bore polythene tee compression-type	0.76
Waste on tee, 5%	0.04
Plumber fixing ¼ hour at £6.00 per hour	1.50
Mate ¼ hour at £4.00 per hour	1.00
	3.30

Unit rate = £3.30 each (labour £2.50, materials £0.80)

15 mm outside diameter soft copper tubing to comply with BS 2871 Table Y, as water service pipe, laid in trench with capillary type fittings. Per linear metre

	£
Consider 30 m of tubing.	
30 m of 15 mm soft copper tube at £85.00 per 100 m	25.50
Waste on copper tube, 5%	1.28
Soft copper tube is marketed in 20 m coils, therefore 1 straight connector is required per 30 m.	
1 No. 15 mm capillary-type straight connector at 50p	0.50
Waste on connectors, 5%	0.03
It will take a plumber and mate about 1 hour to offload and lay 30 m of copper tube in a trench plus about 10 minutes to sweat the joint on the connector.	
Plumber 1 hour 10 minutes at £6.00 per hour	7.00
Mate 1 hour 10 minutes at £4.00 per hour	4.67
	38.98

$$\text{Cost per linear metre} = \frac{£38.98}{30} = £1.30$$

Unit rate = £1.30 per m (labour £0.39, materials £0.91)

15 mm outside diameter copper tubing to comply with BS 2871 Table X, as water-service pipe secured to brick walls with saddle band clips and capillary type fittings. Per linear metre

	£
Consider 30 of tubing.	
30 m of 15 mm copper tubing at £75.00 per 100 m	22.50
Waste on copper tube, 5%	1.13
If 30 m of copper tubing were fixed to walls in one length, 6 straight connectors would be required assuming that copper tubes are marketed in 5 m lengths. In practice, pipe runs are divided into shorter lengths to accommodate the intervening fittings. An allowance of say 3 straight connectors would be reasonable.	
3 straight connectors at 20p each	0.60
Waste on connectors, 5%	0.03
Assume saddle band clips every 1 m, therefore 30 clips required at 5p each	1.50
Waste on clips, 10%	0.15
60 screws at £1.50 per 100	0.90
Waste on screws, 10%	0.09
60 rawlplugs at £1.00 per 100	0.60
Waste on rawlplugs, 10%	0.06
c/fwd	£27.56

		£
	b/fwd	27.56
Labour outputs vary considerably. 30 m of tube in one continuous length could be fixed with an output of 6 m per hour, but as most plumbing installations have tubing divided up into short lengths because of fittings, an average output for this work is about 3 m per hour.		
10 hours plumber at £6.00 per hour		60.00
10 hours mate at £4.00 per hour		40.00
		127.56

$$\text{Cost per linear metre} = \frac{£127.56}{30} = \underline{£4.25}$$

Unit rate = £4.25 per m (labour £3.33, materials £0.92)

Extra for 15 mm capillary copper tee. Each

	£
Cost of 15 mm capillary copper tee, 50p	0.50
Waste on tee, 5%	0.03
It will take a plumber about 5 mintues to cut and clean the end of a copper tube and sweat the capillary joint. 3 ends to one tee, therefore ¼ hour plumber at £6.00 per hour	1.50
¼ hour mate at £4.00 per hour	1.00
	3.03

Unit rate = £3.03 each (labour £2.50, materials £0.53)

Extra for 15 mm compression tee. Each

	£
Cost of 15 mm compression tee, £1.00	1.00
Waste on tee, 5%	0.05
It will take a plumber about 10 minutes to fix a compression tee of this size.	
10 minutes plumber at £6.00 per hour	1.00
10 minutes mate at £4.00 per hour	0.67
	2.72

Unit rate = £2.72 each (labour £1.67, materials £1.05)

Note: For elbows, tees, etc., the amount of pipe displaced is negligible and can therefore be disregarded.

Extra for 15 mm straight tap connector. Each

	£
Cost to purchase of 15 mm tap connector, 60p	0.60
Waste on connector, 5%	0.03
Plumber's jointing paste for screw joint, say 5p	0.05
Plumber fixing 10 minutes at £6.00 per hour	1.00
Mate 10 minutes at £4.00 per hour	0.67
	2.35

Unit rate = £2.35 each (labour £1.67, materials £0.68)

36 mm (nominal) bore unplasticized polyvinyl chloride tubing as waste pipe to comply with BS 4519 with solvent welded joints plugged and screwed to walls with plastic clips. Per linear metre

This type of rigid tubing is normally marketed in 3 or 6 m lengths. Usually the pipe is divided up into relatively small lengths interrupted by bends, tees, connections, etc., and straight connectors will not be required.

	£
Consider 6 m of tubing.	
6 m of 36 mm UPVC tubing at 60p per m	3.60
Waste on tube, 5%	0.18
Allow for clips at 500 mm centres, therefore 12 clips at 6p each	0.72
Waste on clips, 5%	0.04
Screws, plugs, etc., including waste, say	0.44
Solvent welding solution including waste, say	0.15
It will take a plumber about 2 hours to fix 6 m of UPVC tube including plugging and screwing clips.	
2 hours plumber at £6.00 per hour	12.00
2 hours mate at £4.00 per hour	8.00
	25.13

Cost per linear metre = $\frac{£25.13}{6}$ = £4.18

Unit rate = £4.18 (labour £3.33, materials £0.85)

Note: With this type of work the items of bends, tees, connections, etc., will all be measured separately and their costs, particularly the labour element, will be significant and in practice difficult to apportion between the short lengths of pipe and the enumerated fittings.

36 mm UPVC bottle trap with P outlet and 38 mm seal. Each

	£
36 mm UPVC trap complete	1.60
Nominal waste allowance, 2½%	0.04
Plumber fixing 15 minutes at £6.00 per hour	1.50
Mate 15 minutes at £4.00 per hour	1.00
	4.14

Unit rate = £4.14 each (labour £2.50, materials £1.64)

Fix pedestal lavatory basin complete with waste fitting and pair of pillar taps. Each. (The appliance to be supplied against a prime cost sum)

It will take a plumber and mate about 1½ hours to offload, get into position, assemble and screw a lavatory basin on to a wood floor.

	£
1½ hours plumber at £6.00 per hour	9.00
1½ hours mate at £4.00 per hour	6.00
	15.00

Unit rate = £15.00 each (labour £15.00)

The item included in a bill of quantities for 'add for profit' after the prime cost sum should include for insurance against theft or damage, etc.

Alternatively if the estimator is to include for the supply cost within the priced item, he would have to obtain a current quotation for the specified composite item and then add a nominal 2½ per cent allowance for wastage, etc.

Fix low-level WC suite complete. Each

It will take a plumber and mate about 2 hours to offload, assemble and fix only a low level WC suite.

	£
2 hours plumber at £6.00 per hour	12.00
2 hours mate at £4.00 per hour	8.00
	20.00

Unit rate = £20.00 each (labour £20.00)

Fix 1800 long rectangular bath complete with taps, waste, and vitrolite side and end panels. Each

It will take a plumber and mate about 3 hours to offload, get into position, assemble and fix only bath.

	£
3 hours plumber at £6.00 per hour	18.00
3 hours mate at £4.00 per hour	12.00
	30.00

Unit rate = £30.00 each (labour £30.00)

12 Glazing

The majority of glazing is now carried out by sub-contractors and a builder would normally obtain very competitive quotations for this type of work.

Sheet glass is marketed in three qualities: *ordinary glazing quality* which is used for building work, *selected glazing quality,* and *special selected quality.*

Sheet glass, float glass, plate glass, armour plate glass and fancy glasses are all classified according to their thickness in millimetres.

If glass is purchased in cut sizes, the cost per square metre increases slightly, but the waste is reduced from 10 to 5 per cent, and the labour outputs increase accordingly.

Putty for glazing sheet glass (including waste)

	Per square metre of pane sizes shown				
Type of rebate	n.e. 0.1 sq.m	Over 0.1 n.e. 0.5 sq. m	Over 0.5 n.e. 1 sq.m	Over 1 sq.m	*Per linear metre of rebate*
Wood sash	3.2 kg	2 kg	1 kg	0.5 kg	0.20 kg
Metal sash	4 kg	2.5 kg	1.25 kg	0.75 kg	0.25 kg

3 mm ordinary glazing quality clear sheet glass, glazed to wood rebates with linseed oil putty, in panes not exceeding 0.1 square metres. Per square metre

A glazier who is usually a single plumber or painter will cut to size and glaze about 1 m² of glass in this pane classification per 1¼ hours.

	£
1 m² of 3 mm glass at £7.50 per m²	7.50
Offloading and handling glass, say, 15p per m²	0.15
Cutting waste and risk on glass, 10%	0.77
Putty for glazing 3.2 kg at 40p per kg	1.28
Sprigs including waste, say	0.04
Glazier 1¼ hours at £5.00 per hour	6.25
	15.99

Unit rate = £15.99 per m² (labour £6.25, materials £9.74)

3 mm ordinary glazing quality clear sheet glass glazed to wood rebates with linseed oil putty, in panes over 0.1 and not exceeding 0.5 square metres. Per square metre

	£
A glazier will glaze about one m² in one hour, having to cut to size for this pane classification.	
1 m² of 3 mm glass offloaded as before	7.65
Cutting waste and risk on glass, 10%	0.77
Putty for glazing 2 kg at 40p per kg	0.80
Sprigs, say	0.04
Glazier 1 hour at £5.00 per hour	5.00
	14.26

Unit rate = £14.26 per m² (labour £5.00, materials £9.26)

3 mm ordinary glazing quality clear sheet glass glazed to metal rebates with metal casement putty, in panes over 0.1 and not exceeding 0.5 square metres. Per square metre

	£
When glazing panes of this size to metal rebates the output will be slightly less than in glazing to wood. A glazier will glaze about 1 m² in 1¼ hours, including cutting to size.	
1 m² of 3 mm glass offloaded as before	7.65
Cutting waste and risk on glass, 10%	0.77
Metal casement putty for glazing 2.5 kg at 50p per kg	1.25
Glazier 1¼ hours at £5.00 per hour	6.25
	15.92

Unit rate = £15.92 per m² (labour £6.25, materials £9.67)

6 mm thick float glass glazed to wood rebates with loose screwed beads in panes over 0.1 but not exceeding 0.5 square metres. Per square metre

	£
A glazier will glaze with screwed beads about 1 m² of float glass within this pane classification in one hour. (Although the glass is heavier than the previous item, the labour time is reduced because a joiner would normally fix the timber beads.)	
1 m² of 6 mm float glass at £15.00 per m² (including offloading)	15.00
Risk waste only as glass assumed already cut to size, 5%	0.75
Sprigs and back putty, say	0.25
Glazier 1 hour at £5.00 per hour	5.00
	21.00

Unit rate = £21.00 per m² (labour £5.00, materials £16.00)

13 Painting and decorating

Sundry materials: average quantities required per 100 square metres

Varnish on woodwork	6.50 litres
Oil stain on woodwork	5.90 litres
Spirit stain on woodwork	3.64 litres
Cement paint on brickwork	20.32 kg
Creosote and wood preservatives	
1 on sawn wood	15.56 litres
2 on wrot wood	10.92 litres
Tar on woodwork	24.57 litres
Knotting	0.75 litres
Stopping	2.5 kg

Prepare and apply two coats of oil-bound distemper to insulation board ceilings. Per square metre

Consider an area of 100 square metres. First coat will give an average coverage of 220 m² per 50 kg.

Second coat will give an average coverage of 270 m² per 50 kg.

First-coat quantity for 100 m²

$$= \frac{100}{220} \times 50 \text{ kg} = 22.73 \text{ kg}$$

Second-coat quantity for 100 m²

$$= \frac{100}{270} \times 50 \text{ kg} = 18.52 \text{ kg}$$

	£
Total quantity of material, say, 41.25 kg	
41.25 kg of oil bound distemper at £20.00 per 50 kg less 25% trade discount, 41.25 kg at £15.00 per 50 kg	12.38
Residue waste on distemper, 10%	1.24
Applying distemper to insulation board ceilings, a painter will prepare and cover 100 m² in about 10 hours for each coat, i.e. about 10 m² per coat per hour. Therefore 2 coats will require 20 hours.	
c/fwd	£13.62

Average spreading capacities of paints in square metres

	Two-coat treatment				*Three-coat treatment*				
	Washable water paint		*Emulsion paint*		*Primer undercoat*		*Finishing coats*		
Surfaces	1st coat per 50 kg	2nd coat per 50kg	1st coat per 5 litres	2nd coat per 5 litres	Per 5 litres	Per 5 litres	Flat oil or flat finish per 5 litres	Semi gloss or egg shell per 5 litres	Gloss finish per 5 litres
Hard wall plaster	335 – 355	375 – 395	60 – 75	70 – 75	50 – 55	60 – 65	65 – 75	65 – 75	60 – 65
Lime plaster	315 – 335	375 – 395	55 – 60	70 – 75	50 – 55	60 – 65	65 – 75	65 – 75	60 – 65
Hardboard	335 – 355	375 – 420	60 – 70	75 – 80	50 – 55	60 – 65	70 – 75	70 – 75	60 – 65
Insulation board	210 – 230	230 – 315	30 – 35	40 – 45	18 – 22	40 – 45	40 – 45	45 – 50	45 – 50
Concrete and fair face brickwork	210 – 230	250 – 270	40 – 45	50 – 55	35 – 40	40 – 45	45 – 50	45 – 50	45 – 50
Asbestos sheeting	250 – 270	315 – 335	45 – 50	60 – 65	45 – 50	45 – 50	65 – 70	65 – 70	60 – 65
Cement render	250 – 270	290 – 315	45 – 50	55 – 60	40 – 45	50 – 55	50 – 55	50 – 55	50 – 55
Woodwork	—	—	—	—	45 – 55	60 – 65	65 – 75	65 – 75	60 – 65
Steelwork	—	—	—	—	55 – 60	55 – 60	70 – 75	70 – 75	65 – 70

		£
	b/fwd	13.62
20 hours painter at £5.00 per hour including clothing money		100.00
Brushes are supplied by the employer and are very expensive. Brush waste is strictly proportional to the time spent painting and based on present-day wage rate and brush prices, is equal to about 5% of the labour cost.		
Brush waste, 5%		5.00
		118.62

$$\text{Cost per square metre} = \frac{£118.62}{100} = \underline{£1.19}$$

Unit rate = £1.19 per m² (labour £1.00, materials £0.19)

Prepare and apply two coats of oil-bound distemper to plaster walls. Per square metre

Consider an area of 100 square metres of hardwall plaster.

First coat will give an average coverage of 345 m² per 50 kg.

Second coat will give an average coverage of 385 m² per 50 kg

First-coat quantity for 100 mm²

$$= \frac{100}{345} \times 50 \text{ kg} = 14.50 \text{ kg}$$

Second-coat quantity for 100 m²

$$= \frac{100}{385} \times 50 \text{ kg} = 13.00 \text{ kg}$$

	£
Total quantity of material = 27.50 kg	
27.50 kg of distemper at £15.00 per 50 kg	8.25
Residue waste, 10%	0.83
Output of painter distempering walls will be about 7 hours for each coat, i.e. about 14 m² per coat per hour. Therefore 2 coats will require 14 hours painter.	
14 hours painter at £5.00 per hour including clothing money	70.00
Add 5% of labour cost for brush waste	3.50
	82.58

Cost per square metre $= \frac{£82.58}{100} = \underline{£0.83}$

Unit rate = £0.83 per m² (labour £0.70, materials £0.13)

Prepare and apply two coats of Snowcem cement paint to fair face brickwork. Per square metre

	£
Consider an area of 100 square metres.	
Quantity of material for 100 m²:1 coat = 20.32 kg therefore 2 coats will require 40.64 kg at £12.00 per 50 kg net	9.75
Residue waste, 10%	0.98
Output of painter applying cement paint to brick walls will be about 6 hours per coat, i.e. about 16½ m² per coat per hour. Therefore 2 coats will require 12 hours.	
12 hours painter at £5.00 per hour including clothing money	60.00
Add 5% of labour cost for brush waste	3.00
	73.73

Cost per square metre $= \frac{£73.73}{100} = \underline{£0.74}$

Unit rate = £0.74 per m² (labour £0.60, materials £0.14)

Prepare and apply two coats of plastic emulsion paint to plaster walls. Per square metre

Consider an area of 100 square metres of hardwall plaster.
First coat will give an average coverage of 68 m² per 5 litres.
Second coat will give an average coverage of 72 m² per 5 litres.
First-coat quantity for 100 m²

$= \frac{100}{68} \times 5 = 7.35$ litres

Second-coat quantity for 100 m²

$= \frac{100}{72} \times 5 = 6.95$ litres

	£
Total quantity of material = 14.30 litres	
14.30 litres of emulsion paint at £10.00 per 5 litres less 25% trade discount	21.45
Residue waste, 10%	2.15
Output of painter applying emulsion paint to plaster walls will be about 8½ hours per coat, i.e. about 12 m² per coat per hour. Therefore 2 coats will require 17 hours.	
17 hours painter at £5.00 per hour including clothing money	85.00
Add 5% of labour cost for brush waste	4.25
	112.85

Cost per square metre $= \dfrac{£112.85}{100} = \underline{£1.13}$

Unit rate = £1.13 per m² (labour £0.85, materials £0.28)

Prepare and apply one coat plaster primer, one flat oil undercoat and one coat gloss-finish oil paint to plaster walls. Per square metre

	£
Consider an area of 100 square metres of hardwall plaster.	
Plaster primer will give an average coverage of 50 m² per 5 litres	
Primer $= \dfrac{100}{50} \times 5$ litres = 10 litres at £14.00 per 5 litres less 25% trade discount	21.00
Flat oil undercoat will give an average coverage of 60 m² per 5 litres	
Undercoat $= \dfrac{100}{60} \times 5$ litres = 8.33 litres at £13.00 per 5 litres less 25% trade discount	16.24
Gloss-finish paint will give an average coverage of 60 m² per 5 litres	
Gloss finish $= \dfrac{100}{60} \times 5$ litres = 8.33 litres at £14.00 per 5 litres less 25% trade discount	17.49
Total cost of paint	54.73
Residue waste on paints, 10%	5.47
Applying plaster primer to walls will take a painter about 13 hours per 100 m².	
c/fwd	£60.20

		£
	b/fwd	60.20
Applying flat oil undercoat to walls will take a painter about 12 hours per 100 m².		
Applying gloss-finish paint to walls will take a painter about 14 hours per 100 m².		
39 hours painter at £5.00 per hour including clothing money		195.00
Add 5% of labour cost for brush waste		9.75
		264.95

Cost per square metre $= \frac{£264.95}{100} = £2.65$

Unit rate = £2.65 per m² (labour £1.95, materials £0.70)

Prepare and apply one coat of calcium plumbate primer, one flat oil undercoat, and one coat gloss-finish oil paint on general surfaces of steelwork. Per square metre

		£
Consider an area of 100 square metres.		
Metal primer will give an average coverage of 55 m² per 5 litres		
Primer $= \frac{100}{55} \times 5$ litres = 9.11 litres at £16.00 per 5 litres net		29.15
Flat oil undercoat on steel will give an average coverage of 55 m² per 5 litres		
Undercoat $= \frac{100}{55} \times 5$ litres = 9.11 litres at £13.00 per 5 litres less 25% trade discount		17.76
Gloss-finish paint will give an average coverage of 65 m² per 5 litres		
Gloss finish $= \frac{100}{60} \times 5$ litres = 7.7 litres at £14.00 per 5 litres less 25% trade discount		16.17
	Total cost of paint	63.08
Residue waste on paints, 10%		6.31
Applying primer to metalwork will take a painter about 12 hours per 100 m².		
	c/fwd	£69.39

		£
	b/fwd	69.39
Applying flat oil undercoat to metalwork will take a painter about 12 hours per 100 m².		
Applying gloss-finish paint to metalwork will take a painter about 14 hours per 100 m².		
38 hours painter at £5.00 per hour including clothing money		190.00
Add 5% of labour costs for brush waste		9.50
		268.89

Cost per square metre $= \frac{£268.89}{100} = £2.69$

Unit rate = £2.69 per m² (labour £1.90, materials £0.79)

Knot, prime, stop and paint two flat oil undercoats and one coat oil gloss-finish paint on general surfaces of woodwork, externally. Per square metre

		£
Consider an area of 100 square metres.		
On average it will take 0.75 litres of knotting to knot 100 m² of woodwork.		
0.75 litres of knotting at £4.00 per litre net		3.00
On average it will take 2.5 kg of alabastine stopping to stop 100 m² of woodwork.		
2.5 kg of stopping at £2.00 per kg		5.00
Wood primer will give an average coverage of 50 m² per 5 litres		
Primer $= \frac{100}{50} \times 5$ litres = 10 litres at £14.00 per 5 litres less 25% trade discount		21.00
Flat oil undercoat will give an average coverage of 60 m² per 5 litres		
Undercoat $= \frac{100}{60} \times 5$ litres = 8.33 litres per coat, therefore 2 coats will require 16.66 litres at £13.00 per 5 litres less 25% trade discount		32.49
Gloss-finish paint will give an average coverage of 60 m² per 5 litres		
Gloss finish $= \frac{100}{60} \times 5$ litres = 8.33 litres at £14.00 per 5 litres less 25% trade discount		17.49
	Total cost of paint *c/fwd*	£78.98

		£
	b/fwd	78.98
Residue waste on paints and sundry materials, 10%		7.90
Knotting, stopping and applying 1 coat of primer to woodwork will take a painter about 15 hours per 100 m^2.		
Applying flat oil undercoat to woodwork will take a painter about 14 hours per 100 m^2, therefore for 2 coats will require 28 hours.		
Applying gloss-finish paint to woodwork will take a painter about 17 hours.		
60 hours painter at £5.00 per hour including clothing money		300.00
Add 5% of labour cost for brush waste		15.00
		401.88

Cost per square metre $= \dfrac{£401.88}{100} = £4.02$

Unit rate = £4.02 per m^2 (labour £3.00, materials £1.02)

Knot, prime, stop and paint two flat oil undercoats and one coat oil gloss-finish paint on isolated general surfaces of woodwork externally, not exceeding 150 mm girth. Per linear metre

	£
Consider an area of 1 square metre.	
Materials cost of 1m^2 (from previous analysis)	1.02
Labour cost for 1 m^2 (from previous analysis)	3.00
Add 50% to labour cost for 'cutting in' to both edges, etc.	1.50
	5.52

Cost per linear metre $= £5.52 \times \dfrac{150}{1000} = £0.83$

Unit rate = £0.83 per m (labour £0.68, materials £0.15)

Knot, prime, stop and paint two flat oil undercoats and one coat oil gloss-finish paint on isolated general surfaces of woodwork externally over 150 but not exceeding 300 mm girth. Per linear metre

	£
Material cost for 1 m^2 (from previous analysis)	1.02
Labour cost for 1 m^2 (from previous analysis)	3.00
Add 30% to labour cost	1.00
	5.02

$$\text{Cost per linear metre} = £5.02 \times \frac{300}{1000} = \underline{£1.51}$$

Unit rate = £1.51 per m (labour £1.20, materials £0.31)

Prepare and size plaster walls and trim and hang decorative patterned wallpaper PC £5.00 per piece, secured with cellulose paste. Per square metre

	£
Consider 1 roll or piece of wallpaper.	
A roll of English wallpaper is approximately 10 metres long and 0.533 metres wide (i.e. 5.33 m³)	
Cost of 1 roll to purchase ready edged	5.00
Pattern waste, 20%	1.00
Size about 0.05 kg at £1.50 per kg, including waste	0.08
Cellulose paste (allow 5 rolls per packet at £1.00)	0.20
On average a paperhanger will take about 1¼ hours to paste and hang 1 roll or piece of wallpaper including cutting to openings.	
1¼ hours paperhanger at £5.00 per hour (including clothing money)	6.25
	12.53

$$\text{Cost per square metre} = \frac{£12.53}{5.33} = \underline{£2.35}$$

Unit rate = £2.35 per m² (labour £1.17, materials £1.18)

14 Drainage

Excavate trench for pipes not exceeding 200 mm internal diameter in ordinary ground not exceeding 2 m total depth, average 0.50 m starting at ground level. Part backfill with excavated material, remove surplus from site including all necessary earthwork supports, grading and ramming to bottoms. Per linear metre (using hand labour)

For estimating purposes, trenches up to 1 m in depth are taken as being 600 mm wide, 1 m to 2 m deep as 750 mm wide, and over 2 m deep as a minimum of 900 mm wide.

It is assumed that, allowing for the bulking of spoil and the displacement of concrete beds and pipes, about half of the original material is refilled, and half removed from site or subsequently disposed.

Excavation for trenches in ordinary ground not exceeding 2 m in depth (from previous analysis, page 52) = £13.00 per m³.

Returning, filling and ramming excavated material by hand labour (from previous analysis, page 54) = £3.00 per m³.

Removing surplus spoil from site by hand labour, loading and lorry (from previous analysis, page 54) = £10.08 per m³.

	£
Excavating 1 m³	13.00
Refilling ½ m³	1.50
Disposing ½ m³	5.04
	19.54

1 linear m of drain trench 600 mm wide and 0.5 m deep contains 0.30 m³, therefore cost per linear m = £19.54 × 0.30 = £5.86

Grading and ramming to bottoms is not usually priced, and earthwork support is only priced for shallow trenches if site conditions warrant its use.

Unit rate = £5.86 per m (labour £5.86)

Excavate drain trench as before, but average 1.25 m deep. Per linear metre (using hand labour)

1 linear m of drain trench 750 mm wide and 1.25 m deep contains 0.94 m^3 therefore cost per linear metre = £19.54 × 0.94 = £18.37

Unit rate = £18.37 per m (labour £18.37)

Excavate drain trench as before, but average 1.75 m deep. Per linear metre (using hand labour)

1 linear m of drain trench excavation 900 mm wide and 1.75 m deep contains 1.58 m^3 therefore cost = £19.54 × 1.58 = £30.87

Unit rate = £30.87 per m (labour £30.87)

Excavate drain trench as before but over 2 m and not exceeding 4 m total depth, average 2.25 m. Per linear metre (using hand labour)

	£
Excavating 1 m^3 of trench over 2 m but not exceeding 4 m deep by hand from previous analysis (page 52)	20.00
Backfilling ½ m^3	1.50
Disposing ½ m^3	5.04
	26.54

1 linear m of drain trench excavation 900 mm wide and 2.25 m deep contains 1.80 m^3 in the first 2 m depth, therefore cost = £19.54 × 1.80 =	35.17
1 linear m of drain trench excavation 900 mm wide and 2.25 m deep contains 0.23 m^3 in that part of the trench over 2 m in depth, therefore cost = £26.54 × 0.23	6.10
It is assumed here that site conditions warrant the use of a light form of earthwork support; area of trench to be supported = 2/1.00 2.25 = 4.50	
4.5 m^2 at, say, £0.75 per m^2	3.38
	44.65

Unit rate = £44.65 per m (labour £44.65)

Excavate drain trench as before, not exceeding 2 m total depth but average 0.5 m deep. Per linear metre (using ¼ cubic metre excavator)

It is assumed that the excavator will excavate the trenches, throwing the spoil by the side of the open trench, that refilling and ramming will be carried out by hand-labour, and that the machine will later load up surplus spoil into lorries for disposal.

	£
Excavating 1 m³ with machine, trenches not exceeding 2 m deep (from previous analysis, page 49)	2.05
Refilling ½ m³ by hand	1.50
Loading up spoil at rate for reduced level excavation of £1.04 per m³, ½ m³	0.52
Removing spoil from site by lorry from machine, loading at £2.09 per m³ (from previous analysis, page 51) ½ m³	1.05
	5.12

1 linear m of drain trench 600 mm wide and 0.50 m deep contains 0.30 m³, therefore cost per linear metre = £5.12 × 0.30 = £1.54

Unit rate = £1.54 per m (labour £0.81, materials £0.73)

Excavate drain trench as before, but average 1.75 m deep. Per linear metre (using ¼ cubic metre excavator)

1 linear m of drain trench excavation 900 mm wide and 1.75 m deep contains 1.58 m³, therefore cost = £5.12 × 1.58 = £8.09

Unit rate = £8.09 per m (labour £4.25, plant £3.84)

Excavate drain trench as before, over 2 m and not exceeding 4 m total depth, average 2.25 m deep. Per linear metre (using ¼ cubic metre excavator)

	£
Excavating drain trench as before, over 2 m but not exceeding 4 m deep, say	3.00
Refilling ½ m³ by hand-labour	1.50
Loading spoil ½ m³ by machine	0.52
Removing by lorry ½ m³ spoil	1.05
	6.07

	£
1 linear m of drain trench excavation 900 mm wide and 2.25 deep contains 1.80 m³ in the first 2 m depth, therefore cost (as before) = £5.12 × 1.80	9.22
1 linear m of drain trench excavation 900 mm wide and 2.25 m deep contains 0.23 m³ in that part of the trench over 2 m in depth, therefore cost = £6.07 × 0.23	1.40
It is assumed that site conditions warrant the use of a light form of earthwork support 4.5 m² of surface, at, say, £0.75 per m²	3.38
Unit rate = £14.00 per m (labour £7.48, plant £6.52)	14.00

In the previous analysis of rates for drain trench excavation by machine, the banksman has been taken into account in the allocation of the rate between labour and plant. Earthwork support has been included as plant.

Note: In practice it would be possible in many instances to reduce the costs of drainage excavation by increasing the plant usage over all aspects of these items.

Concrete 1:3:6 mix, 38 mm aggregate in 100 mm thick bed, laid under 100 mm internal-diameter drain pipe 450 mm minimum width. Per linear metre

	£
Consider 1 linear m.	
Volume of concrete in the bed = 0.045 m³	
Cost of 1 m³ of 1:3:6 concrete 38 mm aggregate using a 10/7 mixer continuously (from previous analysis, page 63)	39.04
Additional labour in concrete bed to drain pipes, say 2½ hours labourer at £4.00 per hour	10.00
	49.04

Cost = £49.04 × 0.045 = £2.21

Unit rate = £2.21 per m (labour £0.98, materials £1.22, plant £0.01)

Concrete as before in 100 mm thick bed, laid under 150 mm internal-diameter pipe 525 mm minimum width. Per linear metre

Consider 1 linear m.
Volume of concrete in the bed = 0.053 m³
Cost per linear metre = £49.04 × 0.053 = £2.60

Unit rate = £2.60 per m (labour £1.15, materials £1.44, plant £0.01)

Concrete as before in benching to top of 100 mm internal-diameter pipe both sides, including any necessary formwork (bed measured separately). Per linear metre

Consider 1 linear m.
Volume of concrete in the benching with the
pipe volume deducted = 0.015 m^3
Cost of 0.015 m^3 = £49.06 × 0.015 = £0.74

Unit rate = £0.74 per m (labour £0.33, materials £0.40, plant £0.01)

Concrete as before in covering 100 mm internal-diameter 150 mm thick including all necessary formwork (bed measured separately). Per linear metre

Consider 1 linear m.
Volume of concrete in the surround with the
pipe volume deducted = 0.06 m^3
Cost of 0.06 m^3 = £49.04 × 0.06 = £2.94

Unit rate = £2.94 per m (labour £1.31, materials £1.62, plant £0.01)

Approved granular filling in 100 mm thick bed laid under 150 mm internal-diameter drainage pipes, 525 mm minimum width. Per linear metre

	£
Consider 1 linear metre.	
Volume of granular filling in the bed = 0.053 m^3	
Cost of 1 m^3 granular filling unloaded adjacent to trenches, say	15.00
Allow for consolidation and waste, say 33⅓%	5.00
Additional labour in spreading and consolidating filling in trench, say, 2 hours labourer at £4.00 per hour	8.00
	28.00

Cost = £28.00 × 0.053 = £1.48

Unit rate = £1.48 per m (labour £0.42, materials £1.06)

100 mm internal-diameter BS tested quality glazed vitrified stoneware socketed drain pipes to comply with BS 65 and BS 540, laid in trench on concrete, jointed with tarred gaskin and cement mortar 1:3 mix. Per linear metre

Consider 30 m of pipe.

	£
30 m of drain pipe at £1.50 per metre list price	45.00
Add percentage plusage to list price for BS tested quality pipes in loads of over 2 tonnes, 50%	22.50
	67.50
Offloading and stacking pipes: a 1-tonne load contains about 105 pipes and will take 2 labourers about ½ hour each to offload. Therefore 50 pipes or 30 m will take ½ hour labourer's time at £4.00 per hour	2.00
Waste on pipes, 5% of £69.50	3.48
30 m of pipe in 0.6 m lengths will require 50 joints, and it will take about 0.03 m³ of mortar to make 50 joints on 100 mm diameter pipes (including waste) at £39.60 per m³ (from previous analysis)	1.19
It will take about 1 kg of tarred gaskin to make 50 joints on 100 mm diameter pipe (including waste) at £2.50 per kg	2.50
It will take a semi-skilled craftsman about 10 hours to lay and joint 30 m of 100 mm diameter pipe in a trench, at, say, £4.50 per hour	45.00
	121.67

Cost per linear metre $= \frac{£121.67}{30} = £4.06$

Unit rate = £4.06 per m (labour £1.50, materials £2.56)

150 mm internal-diameter BS tested quality glazed vitrified stoneware drain pipes to comply with BS 65 and B540 laid in trench on concrete jointed with tarred gaskin and cement mortar 1:3 mix. Per linear metre

	£
Consider 30 m of pipe.	
30 m of 150 mm drain pipe at £3.00 per m list price	90.00
Add percentage plusage to list price for BS tested quality pipes, 50%	45.00
	135.00
Offloading 1 hour labourer at £4.00 per hour	4.00
	139.00
Waste on pipes, 5%	6.95
About 0.05 m³ of mortar will be required to joint 50 joints on a 150 mm diameter pipe, at £39.60 per m³	1.98
It will take about 1½ kg of tarred gaskin to make 50 joints on 150 mm diameter pipes, at £2.50 per kg	3.75
c/fwd	£151.68

		£
	b/fwd	151.68
It will take a drainlayer about 12½ hours to lay and joint 30 m of 150 mm diameter pipe in a trench, at £4.50 per hour		56.25
		207.93

Cost per linear metre $= \dfrac{£207.93}{30} = \underline{£6.93}$

Unit rate = £6.93 per m (labour £1.88, materials £5.05)

Extra for 100 mm bend. Each

	£
Cost of 100 mm bend to purchase	1.50
Deduct pipe displaced by bend, 450 mm at £2.25 per m with plusage	1.01
	0.49
Add waste, 5%	0.02
Extra cost of bend	0.51
Gaskin and mortar for extra joint at £3.69 for 50 joints, say 7p per joint	0.07
Additional labour laying bend and jointing, drainlayer ¼ hour at £4.50 per hour	1.13
	1.71

Unit rate = £1.71 each (labour £1.13, materials £0.58)

Extra for 100 × 100 mm single junction. Each

	£
Cost of joint to purchase	3.00
Deduct pipe displaced by junction, 750 mm at £2.25 per m with plusage	1.69
	1.31
Add waste, 5%	0.07
Extra cost of junction	1.38
Gaskin and mortar for two extra joints at 7p per joint	0.14
Additional labour laying junction and jointing, drainlayer ½ hour at £4.50 per hour	2.25
	3.77

Unit rate = £3.77 each (labour £2.25, materials £1.52)

100 mm internal-diameter glazed vitrified stoneware drain pipes as before but in runs not exceeding 3 metres long. Per liner metre

	£
Consider 30 m of pipe.	
Materials for 30 m of 100 mm drain pipe (from previous analysis)	76.67
Waste on pipes due to cutting will increase from 5% to 10%, therefore add 5% additional waste	3.84
A drainlayer will lay about 2 m of pipe in an hour in short runs, because of the extra cutting.	
Say 15 hours drainlayer at £4.50 per hour	67.50
	148.01

Cost per linear metre $= \dfrac{£148.01}{30} = £4.93$

Unit rate = £4.93 per m (labour £2.25, materials £2.68)

100 mm internal-diameter outlet glazed vitrified stoneware trapped gulley to comply with BS 539, with 225 × 225 mm dishstone and 150 mm diameter loose painted iron grating, including jointing to drain and surrounding with concrete, and all necessary excavation and disposal. Each

	£
Cost of gulley to purchase complete	4.50
Concrete for setting gulley, say 0.05 m³ at £49.04 per m³ (from previous analysis for drain surrounds)	2.45
Material for joint, say 10p	0.10
Extra excavation for gulley, say 0.1 m³ at £19.54 per m³ for hand-labour (from drain trench analysis)	1.95
Drainlayer setting gulley and jointing, say ½ hour at £4.50 per hour	2.25
	11.25

Unit rate = £11.25 each (labour £4.20, materials £7.05)

100 mm internal diameter vitrified clay pipes with plastic flexible couplings laid in trench on granular filling. Per linear metre

	£
Pipes are mainly marketed in 1.6 m lengths. Consider, therefore, 10 pipes each 1.6 m long.	
16 m of 100 mm pipe at £1.40 per m	22.40
Unloading allow 15 minutes labourer at £4.00 per hour	1.00
c/fwd	£23.40

		£
	b/fwd	23.40
Waste on pipes, 5%		1.17
10 couplings at 70p each		7.00
Lubricant, say		0.30
It will take a semi-skilled craftsman about 2½ hours to lay and joint 16 m of this type of pipe, 100 mm diameter		
2½ hours drainlayer at £4.50 per hour		11.25
		43.12

$$\text{Cost per linear metre} = \frac{\pounds 43.12}{16} = \underline{\pounds 2.70}$$

Unit rate = £2.70 per m (labour £0.70, materials £2.00)

100 mm internal-diameter pitch fibre drainpipes to BS 2760 with tapered joints and couplings laid in trench. Per linear metre

	£
Pipes are marketed in lengths between 1.52 and 3.04 m long. Consider, therefore, 10 pipes each 2.5 m long.	
25 m of 100 mm pipe at £2.00 per m	50.00
Waste on pipes to cut to length, 5%	2.50
10 couplings at £1.20 each	12.00
Waste on couplings, 5%	0.60
A pipelayer and a labourer will lay about 12½ m of pipe in an hour.	
2 hours pipelayer at £4.50 per hour	9.00
2 hours labourer at £4.00 per hour	8.00
(This time includes offloading)	82.10

$$\text{Cost per linear metre} = \frac{\pounds 82.10}{25} = \underline{\pounds 3.28}$$

Unit rate = £3.28 per m (labour £0.68, materials £2.60)

Extra for 90-degree sweep bend on 100 mm internal-diameter pitch fibre pipes. Each

		£
Cost of 90-degree bend to purchase		4.70
Deduct pipe displaced, 900 mm at £2.00 per m		1.80
		2.90
1 extra coupling		1.20
	c/fwd	£4.10

	£
b/fwd	4.10
It will take a pipelayer and labourer about ½ hour to lay bend and make joint.	
½ hour pipelayer at £4.50 per hour	2.25
½ hour labourer at £4.00 per hour	2.00
	8.35

Unit rate = £8.35 each (labour £4.25, materials £4.10)

300 mm diameter spun concrete drain pipes to comply with BS 556, with spigot and socket joints, jointed in cement mortar 1:3 mix, laid in trench. Per linear metre.

	£
Concrete pipes are marketed in varying lengths. Consider, therefore, 10 pipes each 2.5 m long.	
25 m of 300 mm diameter concrete drain pipes at £11.00 per linear m, delivered site	275.00
For offloading 10 pipes, use of mobile 2-ton crane 1 hour at £16.00 per hour, inclusive of fuel and driver (if not included in preliminaries)	16.00
Two labourers, 1 hour each assisting with offloading at £4.00 per hour	8.00
	299.00
Waste cutting to lengths, 5%	14.95
0.03 m³ of cement mortar is required to make 10 joints for 300 mm diameter pipes (including waste) at £39.60 per m³	1.19
1½ kg of gaskin will be required for 10 joints to 300 mm diameter pipes (including waste) at £2.50 per kg	3.75
It will take 4 labourers ½ hour each to lower 1 pipe into the trench and lay into position ready for jointing. Therefore 10 pipes will require 20 hours labourer at £4.00 per hour	80.00
It will take a pipelayer about ½ hour to make 1 joint, therefore 10 joints will take 5 hours at £4.50 per hour	22.50
	421.39

$$\text{Cost per linear metre} = \frac{£421.39}{25} = \underline{£16.86}$$

Unit rate = £16.86 per m (labour £4.10, materials £12.12, plant £0.64)

100 mm diameter cast-iron drain pipes to comply with BS 437, laid in trench with spigot and socket joints, jointed with caulked lead. Per linear metre

	£
Assume pipes marketed in 2.7 m lengths, so consider one 2.7 m length.	
2.7 m of 100 mm diameter cast-iron drain pipe at £14.00 per metre delivered site	37.80
Offloading pipe, say 2 men 2½ minutes each, 5 minutes labourer's time at £4.00 per hour	0.33
Waste on cutting pipes to length, 5%	1.90
Gaskin for joint to 100 mm diameter pipe, say 0.1 kg (including waste) at £2.50 per kg	0.25
Lead for caulking joint, 3 kg at £1.00 per kg (including waste)	3.00
Pipelayer and labourer, say 20 minutes each lowering pipe into trench and laying ready for jointing.	
20 minutes pipelayer at £4.50 per hour	1.50
20 minutes labourer at £4.00 per hour	1.33
Plumber and mate caulking one joint will take ½ hour.	
½ hour plumber at £6.00 per hour	3.00
½ hour mate at £4.00 per hour	2.00
	51.11

$$\text{Cost per linear metre} = \frac{£51.11}{2.7} = \underline{£18.93}$$

Unit rate = £18.93 per m (labour £2.90, materials £16.03)

Concrete (21 MN/m²) in 100 mm thick precast or cast in situ cover slab 1115 × 1340 mm reinforced with mesh fabric weighing 2.22 kg per square metre, perforated and rebated for 610 × 457 mm manhole cover and frame. Each

	£
After deducting the opening, the area is approximately 1.225 m², and requires 0.123 m³ of concrete 1:2:4 mix, intermittent use, say £70.00 per m³ = £70.00 × 0.123	8.61
Say 0.63 m² of formwork to the soffit after allowing wall holds at say, £18.00 per m²	11.34
5 linear m of formwork to 100 mm high edge at say £1.50 per m	7.50
1.25 m² of mesh fabric reinforcement at say, £2.00 per m²	2.50
	29.95

Unit rate = £29.95 each (other cover slabs can be priced on the basis of £20.05 per m²)

610 × 457 mm single-seal flat cast-iron manhole cover and frame to BS 497 Grade C, and set frame in cement mortar 1:3 mix and bed cover in grease. Each

	£
Cost of cover to purchase in lots of 50	19.00
Offloading and stacking, 2½ minutes at £4.00	0.17
Labourer distributing cover on site, say 10 minutes at £4.00 per hour	0.67
Mortar for bedding, say 35p	0.35
Bricklayer 10 minutes bedding at £5.00 per hour	0.83
	21.02

Unit rate = £21.02 each (labour £2.02, materials £19.00)

Coated cast-iron manhole step to BS 1247, built into brickwork. Each

	£
Cost of step to purchase	2.80
Waste, 5%	0.14
Bricklayer walling in step as the work proceeds, say 5 minutes at £5.00 per hour	0.42
	3.36

Unit rate = £3.36 each (labour £0.42, materials £2.94)

100 mm diameter half-round salt-glazed stoneware straight channel 750 mm long and bedding in cement mortar 1:3 mix. Each

	£
Cost of channel 750 mm long to purchase	0.90
Waste for breakages, 10%	0.09
Bedding material, say 10p	0.10
Bricklayer cutting to exact length and bedding in bottom of manhole, say 20 minutes at £5.00 per hour	1.67
	2.76

Unit rate = £2.76 each (labour £1.67, materials £1.09)

100 mm diameter three-quarter-section salt-glazed stoneware branch channel 300 mm long bedded in cement mortar 1:3 mix. Each

	£
Cost of bend to purchase	1.80
Waste for breakages, 10%	0.18
Mortar for bedding, say 10p	0.10
Bricklayer placing in position in benching, say ½ hour at £5.00 per hour	2.50
	4.58

Unit rate = £4.58 each (labour £2.50, materials £2.08)

15 Repairs and rehabilitation work

At the present time it is an established economic fact that existing buildings must be improved and repaired wherever possible in preference to demolition and re-building. This type of work is now very common and is an important part of the work load of the building industry. In many instances, tendering for such work is by means of bills of quantities which require special consideration by the estimator. Generally this type of work is more labour-intensive than new work because:

1 Fewer materials can be obtained as factory-built components
2 More work must be be carried out *in situ* due to the obvious difficulties of some trades working in restricted spaces
3 Manual labour is often necessary instead of mechanical means

These factors normally increase the element of supervision and overhead charges.

In addition to the standard items which are similar to those associated with new construction work, many composite items will be encountered. Work of partial demolition, stripping out, forming or enlarging openings, building up existing openings, builders' work in connection with new specialist services and making good finishings are some examples. An estimator must take into consideration special factors such as temporary protection, security, haulage and tipping fees for demolished materials, and excessive extra handling of new materials. In addition, as materials are often only required in small quantities the supply and haulage costs can be high.

Form new door opening nominal size 1 × 2 metres in existing one-brick thick wall plastered both sides. Provide and build in new concrete lintel size 225 × 150 mm reinforced with two 13 mm diameter mild steel bars, square up jambs, make good plaster and tiled skirting both sides and tiled flooring to threshold all to match existing. (New door, frame, etc., measured elsewhere.) Each

Before the estimator can build up a price for this item he must consider access, the method of removal of rubble, haulage, etc., but most of all he will be influenced by the quantity and position of such work as this will be a key factor for his labour costs. If there is only one other similar type of item in a building it will be inconvenient and very expensive to bring operatives to the site to specially carry out this work. On the other hand if there is a reasonable degree of repetition of this type of work, an efficient use of labour should be possible. In building up the following costs the latter conditions have been assumed. Complete re-decoration later is usually necessary for rooms affected by alterations and therefore making good decorations is not usually specified.

It is always necessary for the estimator to visit the site and identify the exact location and nature of alteration works. Composite items of this nature are often referred to as 'spot items'.

	£
One precast lintel approximately 1.3 m long.	9.00
Allow, say 30 new common bricks to supplement the existing salvaged and cleaned bricks.	
30 bricks at £65.00 per thousand	1.95
Mortar (1:1:6 mix) allow 0.10 m^3 at, say, £40 per m^3	4.00
Plastering materials for both coats, say	5.00
Floor tiles, say	4.50
Bricklayer cutting opening, inserting lintel, squaring and building up jambs, etc. 10 hours at £5.00 per hour	50.00
Attendant labourer, mixing mortar, removing debris, cleaning existing bricks, etc. 10 hours at £4.00 per hour	40.00
Plasterer cutting back, two coat work including making good to existing 4 hours at £5.00 per hour	20.00
Attendant labourer 4 hours at £4.00 per hour	16.00
Floor-tiler, matching floor in new threshold, cutting back and later making good skirting.	
3 hours at £5.00 per hour	15.00
Hire of telescopic props and use of sundry minor timber, say	2.50
Proportional use of lorry, say 1½ hours at £8.36 per hour (see previous analysis)	12.54
Tipping charge, say	6.00
Add a contingency of, say, 10% to cover unforeseen costs and expenses which could be caused by bricks, mortar or plaster proving unusually hard or unusually brittle, or possible difficulties in connection with the sequence of various tradesmen, etc.	18.65
	205.14

Unit rate = £205.14 each (labour £155.10, materials £26.90, plant £23.14)

Take out existing door size 762 × 1981 × 44 mm thick together with casing, architrave, etc. and prepare opening for building up. Each

	£
No materials required.	
Joiner unscrewing door, carefully removing casing and architraves, etc.	
1 hour at £5.00 per hour	5.00
Attendant labourer removing materials to lorry, etc.	
1 hour at £4.00	4.00
Allow a proportional cost of haulage, tipping fees, etc., say	2.00
	11.00

Unit rate = £11.00 each (labour £9.00, materials nil, plant £2.00)

A credit value for the demolished materials must be considered but in most cases there is no such value unless the builder is involved in the selling of demolished materials. Apart from a few materials, e.g. hardcore, slates, lead, etc., items stripped out become an expensive liability because of haulage and tipping costs.

Block up existing opening nominal size 1 × 2 metres in one-brick-thick wall. Make good plaster and skirting both sides to match existing. Each

	£
Common bricks required for 2 m² of one brick wall = 236.	
Allow 260 bricks which will include for the extra materials in toothing and waste, etc.	
260 commons at £65.00 per thousand	16.90
Gauged mortar 0.12 m³ at £40.00 per m³	4.80
Plastering materials for two coat work, say	5.00
2 lengths of skirting to match existing, say	3.50
	30.20

A bricklayer's output under these conditions will fall considerably to 20 – 40 bricks an hour depending on the amount of work available. For this particular operation allow 12 hours which will take into consideration the extra work involved in cutting and toothing to the jambs of the opening and wedging and pinning to the soffit.

c/fwd £30.20

		£
	b/fwd	30.20
Bricklayer 12 hours at £5.00 per hour		60.00
Attendant labourer 6 hours at £4.00 per hour		24.00
A plasterer will normally apply 1 m^2 of each coat of plaster in about 15 minutes, but allow 1 hour for each side of the opening per coat for cutting back, making good and generally working in relatively small areas 1 hour × 2 sides × 2 coats = 4 hours.		
Plasterer 4 hours at £5.00 per hour		20.00
Attendant labourer 4 hours at £4.00 per hour		16.00
A joiner will take approximately 1½ hours to cut, fix and make good the skirtings to both sides of the built up opening. If the mouldings of the existing skirting are non-standard, this time allowance would have to be increased for the *in situ* work necessary in matching.		
Joiner 1½ hours at £5.00 per hour		7.50
		157.70

Unit rate = £157.70 (labour £127.50, materials £30.20)

16 Civil engineering work

Estimating and tendering for civil engineering work is generally more complicated and requires considerable experience. The main points to bear in mind are as follows:

1 The principle of measurement and presentation in bills of quantities are different from building work. Items are described with very little detail and have to be priced bearing in mind the scope of the method of measurement
2 There are three different main methods of measurement:
   The SMM for Civil Engineering Quantities (1972 amended edition)
   The 1976 Civil Engineering SMM
   The DoE Method of Measurement for Roads and Bridges
3 The work involved is generally of a different nature and more uncertain in extent than building work
4 The rates to be assessed for items will probably have to include for work which is not necessarily measured
5 Due to the nature of civil engineering, some projects may require substantial temporary works. On many occasions this is left to the estimator who will have to arrange for the design, planning, etc., before he can price the work involved. Such work may not be itemized specifically in the tender documents
6 Certain items may be measured at the discretion of the quantity surveyor, for example, formwork, working space, etc. This possibility requires an estimator to exercise special care that all necessary costs have been included in the tender
7 The plant used is likely to be of a more involved nature and cost than for building work, depending on the type of contract
8 Usually both the detailed drawings and the specification are of high significance to the estimator because they contain much information that is not included in the bills of quantities

A simple typical item which illustrates the basic problem for an estimator would be tunnel manhole shaft excavation. The item may merely state 'excavation for manholes not exceeding 6 metres deep – 105 cubic metres'. It will be necessary for the estimator to study the drawings and specification and to consider a typical manhole and the method of construction. He will himself design the excavation that he considers will be actually necessary and compare this with the approximate volume for one manhole as shown in the bill of quantities together with working space items (if measured).

He then calculates the actual operational costs involved for the excavation, temporary supports, pumping, trimming, backfilling, disposal (including haulage and tipping charges), guard rails, staging, hoisting, etc. for the excavation work to this manhole. This cost will include all labour, plant, machinery, fuel, overheads and profit although some contractors may wish to allocate some of these charges to preliminary items elsewhere. Assume that the above operations cost an estimated total of £850.00 for the one typical manhole and that the 105 m³ in the bills of quantities represents five similar manholes. This would give a measurement 'allowance' for one manhole of 21 m³ and a unit rate to be inserted of

$$\frac{£850.00}{21} = £40.48 \text{ per m}^3.$$

It must be emphasized that this is a simple example and when the scope and range of all civil engineering is considered the difficulties of tendering for such work can be appreciated.

17 Daywork

Daywork is a means of valuing and agreeing the cost of work carried out under certain circumstances and is based on the actual cost of labour, materials and plant used plus overheads and profit. As far as the Standard Forms of Contract issued by the Joint Contracts Tribunal are concerned, the rules for establishing the basis of valuation of daywork costs are contained in the 'blue' booklet, *Definition of prime cost of daywork carried out under a building contract,* published jointly by the RICS and NFBTE.

Prior to 1966, contractors tendering under JCT Contracts all used the same formula and generally were paid the same rates for work valued on a daywork basis. In 1966 a new formula was agreed and tendering contractors were instructed to fill in their own 'percentage adjustments' to the defined prime cost in accordance with Clause 11 (4)(c) of the JCT Contract.* The 'red' booklet contained the basis and guidance for work of a jobbing nature.

On 1 December 1975 a revised 'blue' book was published and estimators had to revise their calculations by working out the total actual anticipated costs per hour or week including whatever overhead charges and profit were required and deducting from this the new prime cost 'allowances'. The difference is then expressed as a percentage of the prime cost and inserted on the tender documents. This is illustrated as follows:

		Craftsman		**Labourer**
		£		£
Total costs per 40 hour week for the typical 'all-in' rate from the detailed build up illustrated in Chapter 1, say		145.00		120.00
Add Overhead charges, Preliminaries, profit, etc. assume that 35% is required		50.75		42.00
Actual weekly rates required by the contractor	*c/fwd*	£195.75	*c/fwd*	£162.00

*See 1980 edition, Clause 13.5.4.1.

		Craftsman		**Labourer**
		£		£
	b/fwd	195.75	*b/fwd*	162.00
Deduct base allowances as defined in the booklet				
Guaranteed minimum weekly earnings	80.40		68.60	
Employer's National Insurance Contributions at 13.7%	11.01		9.40	
CITB Levy	0.69		0.12	
Annual Holiday Credit Stamps including death benefit scheme and public holidays	8.60	100.70	8.60	86.72
Extra payments to be included in the 'percentage addition'		£95.05		£75.28

Calculation of percentage additions:

$$\text{Craftsman} = \frac{£95.05}{£100.70} \times 100 = 94.39\%$$

$$\text{Labourer} \frac{£75.28}{£86.72} \times 100 = 86.81\%$$

Conclusion: insert a 90% addition on labour

It should be stressed that this example indicates the principles involved and that different contractors will have marginally different conclusions for most of the items based on their own particular circumstances. However the 35 per cent overheads and profit, and the extra bonus payments are both subject to wide variations and will influence the final percentage. In practice an estimator must obtain all the necessary up-to-date information relevant to his employer and operatives and apply the principles defined in the booklet. If a contractor decided that he required, for example, a weekly 'all-in' rate of £250.00 for a craftsman due to his view of the possible disruptive effects of variations carried out at daywork rates, the addition would be about 150 per cent. On the information and rates of pay, etc., ruling at the date of publication of this book, the authors consider a labour percentage of 150 per cent to be excessive unless there were exceptional justifying circumstances but nevertheless a contractor is free to insert whatever amount he wishes, although sometimes these daywork costs are reflected within the tenders.

The percentage additions for both materials and plant are to include for a proportion of overheads, administration and profit and figures in the range of 10 to 20 per cent would be considered reasonable under normal conditions. Care must be taken for the plant percentages to be applied on current hire charges or up-dated standard hire charges.

It is possible to agree other methods of assessing the costs of daywork. Examples are for work carried out during the Defects Liability Period of a JCT Contractor or for civil engineering work but estimators should carefully read the instructions to tenderers and consider the financial implications before entering any percentage additions on a tender document.